T0211918

SpringerBriefs in Energy

SpringerBriefs in Energy presents concise summaries of cutting-edge research and practical applications in all aspects of Energy. Featuring compact volumes of 50 to 125 pages, the series covers a range of content from professional to academic. Typical topics might include:

- A snapshot of a hot or emerging topic
- A contextual literature review
- A timely report of state-of-the art analytical techniques
- An in-depth case study
- A presentation of core concepts that students must understand in order to make independent contributions.

Briefs allow authors to present their ideas and readers to absorb them with minimal time investment.

Briefs will be published as part of Springer's eBook collection, with millions of users worldwide. In addition, Briefs will be available for individual print and electronic purchase. Briefs are characterized by fast, global electronic dissemination, standard publishing contracts, easy-to-use manuscript preparation and formatting guidelines, and expedited production schedules. We aim for publication 8–12 weeks after acceptance.

Both solicited and unsolicited manuscripts are considered for publication in this series. Briefs can also arise from the scale up of a planned chapter. Instead of simply contributing to an edited volume, the author gets an authored book with the space necessary to provide more data, fundamentals and background on the subject, methodology, future outlook, etc.

SpringerBriefs in Energy contains a distinct subseries focusing on Energy Analysis and edited by Charles Hall, State University of New York. Books for this subseries will emphasize quantitative accounting of energy use and availability, including the potential and limitations of new technologies in terms of energy returned on energy invested.

More information about this series at https://link.springer.com/bookseries/8903

Jay Doorga · Soonil Rughooputh ·
Ravindra Boojhawon

Geospatial Optimization of Solar Energy

Cases from Around the World

 Springer

Jay Doorga (iD)
Faculty of Sustainable Development &
Engineering, Department of Emerging
Technologies
Université des Mascareignes
Rose Hill, Mauritius

Soonil Rughooputh
Faculty of Science, Department of Physics
University of Mauritius
Reduit, Mauritius

Ravindra Boojhawon
Faculty of Science, Department
of Mathematics
University of Mauritius
Reduit, Mauritius

ISSN 2191-5520   ISSN 2191-5539 (electronic)
SpringerBriefs in Energy
ISBN 978-3-030-95212-9   ISBN 978-3-030-95213-6 (eBook)
https://doi.org/10.1007/978-3-030-95213-6

This Springer imprint is published by the registered company Springer Nature Switzerland AG
The registered company address is: Gewerbestrasse 11, 6330 Cham, Switzerland

In Memory of Mr Beeshamsingh Doorga,
now in Heaven

Foreword

The costs of new renewable energy sources have fallen dramatically in recent years. It is increasingly clear that renewables will play the critical role in future electricity systems, as the global effort for decarbonization gathers pace. Moreover, the global solar resource vastly exceeds humanity's demand for energy. And this resource is best in the countries within, and close to, the tropics. So solar electricity will be critical to the sustainable development of the growing economies in these countries.

But knowing that solar energy will be important is only part of what we need to understand. There are other questions. What is the relative importance of rooftop installations and solar farms? And how should the latter be located, given the important constraints, such as load proximity, network connection, topography, ecological impacts and other options for use of the same land?

The authors of this book have produced an excellent manual for answering these questions. Starting from the fundamental science, and using state-of-the-art geographical information systems' approaches for resource assessment, they systematically address the key issues. The result is an invaluable guide for decision-makers seeking to make best use of solar energy, and an excellent resource for everyone interested in the challenge of deploying the world's most important renewable resource.

October 2021

Prof. Nick Eyre
Environmental Change Institute
University of Oxford
Oxford, UK

Preface

The increasing energy demand stemming from rapid urbanization and economic development is causing serious strain on the dwindling reserves of fossil fuels. With the alarming rise in global temperature since pre-industrial levels, the global community is calling for immediate sustainable actions. Blessed with the abundance of solar resources, mostly concentrated in the tropical and sub-tropical regions, countries worldwide are tapping into this energy source. The integration of solar energy in the global energy mix is being accelerated by the significant decline in the cost of photovoltaic modules over the years. However, the implementation of photovoltaic facilities necessitate substantial land resources, which is becoming scarce with population growth, competing land uses, and climate change.

This brief provides an overview on how countries can sustainably plan their land resources so as to maximize solar energy yield whilst minimizing land space. Three main themes are covered: Modelling the solar radiation climate to optimize on land resources; Determining best regions for utility-scale solar facilities; Exploitation of available rooftop space to dampen land requirements. The policy implications that would help countries boost their solar energy markets are also elaborated.

This document has been compiled with the hope that it may serve as a reference to provide an informed decision to all those who are concerned with the enthusiastic task of promoting sustainable development through the efficient harnessing of the solar resource. The organization of this document is in such a way as to provide a comprehensive knowledge and guide on how to exploit the solar energy resource to a wide audience, ranging from researchers to policy makers and people in general.

Rose Hill, Mauritius
October 2021

Dr. Jay Doorga
Prof. Soonil Rughooputh
Assoc. Prof. Ravindra Boojhawon

Acknowledgements

We would like to acknowledge the support of Angeliki Athanasopoulou, Cecil Joselin Simon and Cynthia Kroonen for coordinating this book project. The completion of this work would not have been possible without the inputs from NASA Earth Observation, Mauritius Meteorological Services, Revue Agricole et Sucrière de l'Île Maurice, OpenStreetMap, Ministry of Housing and Lands (Cartographic Section), and Statistics Mauritius. We extend our thanks to Ryan Tannoo for the architectural layout of the Solar City design proposed. The facilities and resources provided by the Université des Mascareignes and the University of Mauritius are also gratefully acknowledged.

Contents

About the Authors

Dr. Jay Doorga is a lecturer in the Department of Emerging Technologies, Faculty of Sustainable Development & Engineering, Université des Mascareignes. He holds a M.Sc. in Environmental Change and Management from the University of Oxford and a Ph.D. in GIS: Modelling and Forecasting from the University of Mauritius. He is the recipient of academic honours and awards, including the Oxford ECM Dissertation Award, Cambridge Outstanding Achievement Award, State of Mauritius Laureate Scholarship and Gold Medalist of the University of Mauritius. His research interests focus on exploiting renewable energy resources and mitigating the impacts of climate change using Geographical Information Systems (GIS). At the Mauritius Oceanography Institute, he has led a national research project funded by the Government of Mauritius to explore the wave energy potential of Mauritius. He holds a patent for the invention of a wave energy prototype.

Prof. Soonil Rughooputh is a Professor in the Department of Physics, Faculty of Science, University of Mauritius since 1999. He has a Ph.D. in Physics from the University of London and conducted research at the Davy Faraday Laboratory at The Royal Institution of Great Britain. He was the first CEO of the Mauritius Renewable Energy Agency and has held various positions at the University of Mauritius; Acting Vice-Chancellor, Pro Vice-Chancellor, Dean and Head of the Department of Physics. He was a Visiting Scientist and Post-Doctoral Research Fellow at the University of California, Santa Barbara, USA, and he has worked in California as a Research Scientist. He also worked for Physical Optics Coorporation and Wyatt Technology Inc—both in California, USA. Professor Rughooputh is known for one-dimensional exciton propagation; chromism in conjugated polymers; electronic characterization of conducting polymer gels; and energy and environment modelling. He has contributed to over 180 scientific publications and one patent.

Assoc. Prof. Ravindra Boojhawon is an Associate Professor in the Department of Mathematics, Faculty of Science, University of Mauritius. He has a Ph.D. in the field of Numerical Linear Algebra. He has extensive experience in developing GUIs and programming languages such as MATLAB, R, Microsoft Excel VBA, Mathematica and GIS software. He has also developed expertise over the years in consultancy services where he has used his programming skills and geostatistical background in problem solving related to environmental science. Dr. Boojhawon has over 40 scientific publications in wide-ranging fields from the Environmental Science to Life Science and Mathematics.

Acronyms

AHP	Analytical Hierarchy Process
AM	Air Mass
BESS	Battery Energy Storage System
CAD	Computer-aided design
CLARA-A2	Cloud, Albedo, and Surface Radiation edition 2
CMSAF	Satellite Application Facility on Climate Monitoring
COVID	Coronavirus Disease
CSP	Concentrated Solar Panel
DFHI	Diffused Horizontal Irradiance
DFNI	Diffused Normal Irradiance
DHI	Direct Horizontal Irradiance
DNI	Direct Normal Irradiance
GHI	Global Horizontal Irradiation
GIS	Geographic Information System
IPP	Independent Power Producer
IRENA	International Renewable Energy Agency
LCOE	Levelized Cost of Energy
LiDAR	Light Detection and Ranging
MCDA	Multi Criteria Decision Analysis
NASA	National Aeronautics and Space Administration
NEO	NASA's Earth Observation
NWP	Numerical Weather Prediction
PV	Photovoltaic
RNI	Reflected Normal Irradiance
RPS	Renewable Portfolio Standard
SARAH-E	Surface Solar Radiation-Heliosat-East
SDG	Sustainable Development Goal
SREC	Solar Renewable Energy Certificates
SWH	Solar Water Heater
WLC	Weighted Linear Combination
WRF	Weather Research Forecasting

Chapter 1
Introduction

Abstract Geopolitics and the overexploitation and ensuing depletion of fossil fuel resources to power the global economy have led to energy security issues. This is further compounded by the surplus release of carbon dioxide in the atmosphere, caused by the burning of fossil fuels, and which threatens to disrupt planetary functioning and warm the climate at an unprecedented rate, hopefully not beyond repair. Solar energy is one of the most promising renewable energy resource that can help to decarbonise fossil fuel-reliant power systems worldwide. The economic prospects offered by this tremendous energy resource will revolutionize the global power sector. This chapter presents an overview of the global energy situation and the geospatial constraints involved whilst exploiting the solar energy resource. The scope of the current work is presented, followed by the book structure which details the contents of each chapter.

1.1 Global Energy Context

Global energy security is a key endeavour in the current context of population expansion, rapid economic development, diminishing fossil fuel supply, and political instability in fossil fuel-exporting countries [1]. This situation has been further exacerbated by the COVID pandemic, which left fossil fuel-reliant countries more exposed to economic shocks. The global energy consumption is still dominated by fossil fuels, which accounted for about 84% (33.1% Oil; 27% Coal; 24.3% Gas) of the global energy mix in 2019 [2]. The release of large quantities of carbon dioxide in the atmosphere due to the overconsumption of fossil fuels to power the global economy led to an increase in global temperature which is forced to rise at an unprecedented rate. This temperature rise is having severe repercussions on: (1) the climate level with more frequent and intense extreme weather events; (2) ecological level with long-term physiological and behavioural changes, and high extinction rates in terrestrial and marine organisms; (3) planetary level with an alteration of the biogeochemical cycles. The injection of one trillion tonnes of carbon in the atmosphere, more than half of which has already been released would result in a peak warming of 2 °C, causing irreversible changes to the planetary functioning and the biological organisms inhabiting it [3].

To dampen reliance on fossil fuels, a shift to low-carbon energy sources such as nuclear or renewable energies, is necessary. Renewable energy will be a key factor in the decarbonisation of global energy systems. Renewables accounted for 11.4% of global energy consumption in 2019, with hydropower contributing 6.4%, followed by wind (2.2%) and solar (1.1%) [2]. The global average cost reductions of 69% in the electricity generated from utility-scale solar power plants from \$0.36/kWh to \$0.11/kWh, brought by significant cost reductions in solar modules, paves the way for a solar energy revolution in the incoming decades [4]. It is expected that solar photovoltaic (PV) would generate 25% of the global electricity needs, becoming a major generation source by 2050 [5]. The Levelized Cost Of Electricity (LCOE) of utility-scale solar is falling and is even lower than coal-fired power stations and combined-cycle gas turbines in some high solar resource potential areas around the world. In the upcoming decades, the economic prospects brought by the falling prices of generating electricity from solar PV is expected to outweigh the scenario of maintaining existing fossil fuel-based energy infrastructures. It is therefore inevitable that solar energy will be mankind's most promising source of power to sustainably fuel the economy of tomorrow.

1.2   Solar Resource Potential and Geospatial Constraints

Climate change-induced land degradation and the rise in sea level caused by an increase in surface temperature are spawning a redefinition of land management. Vulnerable communities are being forced to confront and adapt to the impacts of climate change as small islands are facing land constraints with a notable decrease in land area associated with the rising sea level, increased salanisation of the coastal areas, continued damages to coastal infrastructures, and loss of lives of coastal communities due to increased number of storm surges coupled with the degradation of natural coastal defences provided by marine biota. Increased habitat degradation, wildfires and floods worldwide are inspiring more judicious use of land resources due to the precarious nature of future climate-related land degradation. On top of the growing concern for diminishing land space, nations face the common challenge of having to keep pace with the rapidly growing energy demand, which place enormous strain on the dwindling supplies of fossil fuels dominating the global energy mix [6].

Significant global average cost reductions in solar photovoltaics have spurred adoption worldwide as countries seek to diversify their energy mix and enhance the security of supply through the progressive market penetration of photovoltaic systems [7]. However, the competing interest for land space from the solar energy industry conflicts with the urbanization, infrastructure and food production sectors. At the intersection of these two societal priorities to meet energy demand whilst coping with the diminishing and competing land space, lies an area where it is possible to maximize energy need at the requirement of the minimum land space. Consequently,

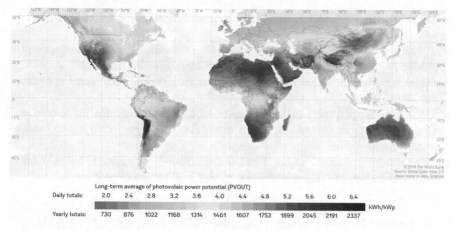

Fig. 1.1 Global photovoltaic power potential (Reused with permission from [9] Open Access under a CC BY 4.0 license, https://creativecommons.org/licenses/by/4.0/, map obtained from the "Global Solar Atlas 2.0", a free, web-based application is developed and operated by the company Solargis s.r.o. on behalf of the World Bank Group, utilizing Solargis data, with funding provided by the Energy Sector Management Assistance Program (ESMAP))

this book seeks to propose a solution to enable countries worldwide to make efficient use of their terrestrial resources in order to maximize energy generation. An estimated share of 50% of the global population is expected to live in the tropics in the late 2030s [8]. The expanding population in tropical and sub-tropical regions combined with the tremendous potential of photovoltaic power potential on spatial scales (Fig. 1.1) urge the adoption of solar energy technologies.

The work presented in this book will help guide countries worldwide to adopt a sustainable path through the strategic investments in solar energy technologies in their quest to dampen reliance on fossil fuel consumption and progressively decarbonise power sectors.

1.3   Scope of This Work

We provide a scientifically robust methodological approach that would permit countries to address the most pressing issues faced by policy makers worldwide regarding the planning and exploitation of solar energy. Besides providing a concise and up-to-date guide on solar resource assessment, this book encompasses wide-ranging benefits, as summarized below:

- A concise theoretical background on the solar energy process is provided in Chap. 2 to equip readers with a knowledge pool in order to comprehend the more advanced research presented in the later stages.

- This work provides a scientifically robust method for countries to optimize, manage and exploit their solar energy resources.
- The methods and tools employed for analysing the solar energy resource potential are based on cutting-edge research that aspires to incorporate legal, geotechnical, environmental, social and climatological factors on a GIS platform to assist decision-making.
- Methodologies proposed are accompanied by case studies worldwide to provide a guide on the application of models discussed for solar energy analysis and assessments.
- The material presented covers a broad area from solar resource assessment to the identification of ideal solar farm sites and exploration of rooftop photovoltaic potential.

This work seeks to disseminate knowledge on the solar resource potential to a wide audience, ranging from architects, urban planners, agro-industrialists, climatologists, policy makers, researchers and energy experts, amongst others.

1.4 Book Structure

The book is structured as follows: Chap. 2 provides a theoretical foundation on solar energy processes from the Sun's core to the Earth's surface. Chapter 3 explores ways of conducting solar resource assessments to guide land planning and management. The methodology adopted to determine optimum sites for utility-scale solar farms is provided in Chap. 4. In Chap. 5, the technique used to perform a rooftop photovoltaic resource assessment is presented. We offer a reflection on the geospatial management of land resources in the energy optimization context and present conclusive remarks in Chap. 6.

References

1. Kowalski G, Vilogorac S (2008) Energy security risks and risk mitigation: an overview. https://unece.org/fileadmin/DAM/oes/nutshell/2008/9_Energy_Security_Risks.pdf. Cited 30 August 2021
2. Ritchie H, Roser M (2020) Overview of global energy. https://ourworldindata.org/energy-overview. Cited 30 August 2021
3. Allen MR, Frame DJ, Huntingford C, Jones CD, Lowe JA, Meinshausen M, Meinshausen N (2009) Warming caused by cumulative carbon emissions towards the trillionth tonne. Nature 458(7242):1163–1166
4. IRENA (2017) Renewable power: sharply falling generation costs. https://www.irena.org/-/media/Files/IRENA/Agency/Publication/2017/Nov/%20IRENA_Sharply_falling_costs_2017.pdf. Cited 31 August 2021
5. IRENA (2019) Future of solar photovoltaic deployment, investment, technology, grid integration and socio-economic aspects. https://www.irena.org/-/media/Files/IRENA/

Agency/Publication/2019/Nov/IRENA_Future_of_Solar_PV_summary_2019.pdf?la=en& hash=A626155A0775CC50427E23E7BE49B1AD2DD31073. Cited 31 August 2021

6. Lefevre N (2010) Measuring the energy security implications of fossil fuel resource concentration. Energy Policy 38(4):1635–1644
7. Pillai U (2015) Drivers of cost reduction in solar photovoltaics. Energy Econ 50:286–293
8. WPR (2021) Tropical countries 2021. https://worldpopulationreview.com/country-rankings/tropical-countries. Cited 14 August 2021
9. Wiki (2017) Photovoltaic power potential. https://commons.wikimedia.org/wiki/File:World_PVOUT_Solar-resource-map_GlobalSolarAtlas_World-Bank-Esmap-Solargis.png. Cited 15 August 2021

Chapter 2
Solar Energy

Abstract This chapter lays the theoretical foundation governing solar radiation processes and variations on both spatial and temporal scales. The origin of the insolation parameter in the Sun's core, its journey to the Earth's atmosphere, the interaction with the atmosphere, and its conversion into electricity in photovoltaic modules are discussed. The spatial and temporal distributions of the insolation parameter arising from seasonal/latitudinal variations and cloud cover effects are explored. The work presented in this chapter aims to provide guidance to knowledge seekers and enthusiasts who aspire to understand the solar energy process.

2.1 Radiation Journey from Sun to Earth

The Sun is the most important driving force of the Earth's climate system and the redistribution of its energy controls the structure and dynamics of the atmosphere and ocean. Nuclear fusion reactions taking place in the thermonuclear core (Fig. 2.1) give rise to interior temperatures reaching about 10^7 K and an interior radiation flux of variable spectral distribution [2]. The rate of hydrogen fusion into helium is high owing to suitable temperatures of the core. Photons are thereafter delivered to the convective layer through the radiative zone and subsequently suffer many interactions as they slowly diffuse outward in a series of 'random walks' until they reach the photosphere. This process of photon movement from the core to the surface of the Sun takes approximately 170,000 years, reflecting on multiple collisions taking place in the Sun's structure [3]. The outer layer has an approximate average temperature of 5800 K and displays a relatively continuous spectral distribution.

Radiation streams out of the Sun at a rate of 3.85×10^{26} W, but only half of a billionth of the total energy is intercepted by the Earth, principally due to the large separation of the order of 150×10^6 km [4]. A solar constant of 1368 W/m^2 has been recorded by satellite measurements and when averaged over the Earth's surface, this value comes down to 342 W/m^2 at the top of the atmosphere. Numerous factors influence and are linked to the radiation received on the Earth's surface, among which lie atmospheric contribution (due to absorption by atmospheric gases), albedo of the Earth's surface components (cloud cover, ice cover, water bodies among others) and

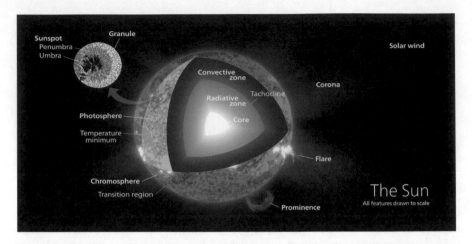

Fig. 2.1 Anatomy of the Sun (Reused with permission from [1] Open Access under a CC BY-SA 3.0 license, https://creativecommons.org/licenses/by-sa/3.0/)

meteorological factors (sunshine duration, air temperature, winds, relative humidity among others). These factors will influence to some extent the variations in both space and time of solar radiation parameter.

The atmospheric composition and dynamics play an important role in determining the geospatial variations of insolation. We investigate in this chapter, the various aspects governing the interaction and distribution of solar radiation on Earth.

2.2 Solar Radiation Budget

Several climate models have been implemented over the years to understand the Earth's solar radiation budget. Figure 2.2 illustrates the processes which alter the incoming insolation as it moves through the atmosphere. Out of the incoming short-wave radiation amounting to 341 W/m^2, about 79 W/m^2 is reflected by the presence of clouds, aerosol and constituents that make up the atmosphere, while about 78 W/m^2 gets absorbed by the atmosphere. An additional 161 W/m^2 is absorbed by the Earth's surface. Radiative balance between incoming and outgoing radiation is established through the emission of longwave terrestrial radiation (as illustrated on the right of Fig. 2.2) and the reflection by the surface amounting to 23 W/m^2. The whole mechanism influences the equilibrium temperature of the Earth-atmosphere system [5].

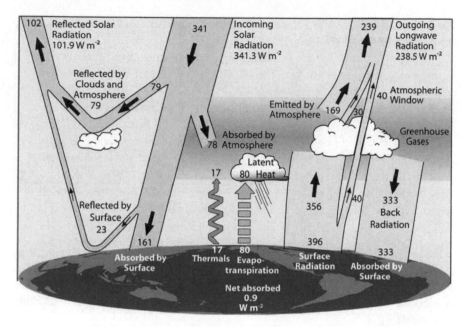

Fig. 2.2 Earth's solar radiation budget (Reused with permission from [6] Copyright (2009) (American Meteorological Society))

2.3 Solar Spectrum Theory and Atmospheric Interaction

Extraterrestrial radiation is the quantity of radiation incident on the uppermost part of the atmosphere, at what is referred to as zero (0) air mass (AM0). The Sun being a blackbody, emits radiation across most of the electromagnetic spectrum, which is defined by its average surface temperature. Ignoring the Sun's age consideration, the radiation detected at the uppermost section of the Earth's atmosphere varies in temporal scales due to the elliptical path of the Earth as it moves around the Sun, and solar activities which comprise of solar flares and sunspots. An increase of 7% in Extraterrestrial solar radiation is observed from July to January, corresponding to the time when the Earth is closest to the Sun [7].

As extraterrestrial radiation penetrates the Earth's atmosphere, it gets attenuated by different extents (due to absorption and scattering processes) depending on the relative concentrations of atmospheric constituents and the optical path length [8]. The major portion of solar radiation that attains the Earth's surface as 'shortwave' radiation is of wavelength shorter than $4.0\,\mu$m whereas that emitted by the Earth lies in the broad waveband from 4.0 to $40.0\,\mu$m and is classified as 'longwave' radiation [9]. The important absorbers of solar radiation within the mixture of gases which makes up the atmosphere are: Oxygen and ozone (absorb radiation less than $0.3\,\mu$m in the uppermost layers of the atmosphere with the remaining percentage back-scattered to outer space); Carbon dioxide (with strong absorption band around $15\,\mu$m and

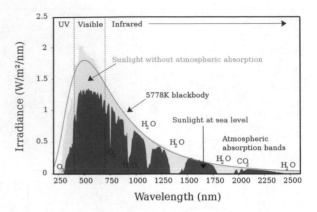

Fig. 2.3 Solar radiation spectra at Earth's surface and outside of the atmosphere (Reused with permission from [11] Open Access under a CC BY-SA 3.0 license, https://creativecommons.org/licenses/by-sa/3.0/)

is the main cause of attenuation in lower stratosphere); water vapour (regarded as the most absorbing atmospheric gas with strong absorption around $6\,\mu m$) [10]. The absorption, reflection and scattering processes cause the curve of solar irradiance incident on the Earth's surface to 'spiky' and fall below that of the extraterrestrial irradiance (Fig. 2.3).

At around $10\,\mu m$ corresponding to Earth's average temperature, lies a small band referred to as the atmospheric window, where there is relatively little absorption of longwave radiation by atmospheric gases. The narrow windows in the shortwave part of the spectrum also permeate a significant portion of the incoming solar radiation, where the majority reaches the Earth's surface. The incoming and outgoing solar radiation that permeate through the windows dictate the Earth-atmosphere energy balance.

2.4   Solar Radiation Components

Extraterrestrial radiation from the Sun is split into several components upon encountering particles in the atmosphere. Part of this radiation originating from the Sun's disk is influenced by the presence of cloud, anthropogenic gases or dust among others resulting in scattering of the radiation. The amount of scattering is dependent on the size of the particles and the wavelength of incoming radiation. Back-scattered radiation would occur in addition to the fact that some of the incoming solar radiation is redirected back to outer atmosphere. Moreover, the process of absorption would cause some gases such as carbon dioxide to selectively absorb solar radiation through the mechanism of the vibration of their molecules [12].

The non-scattered and unabsorbed solar radiation constitute the beam or direct radiation [2]. The combination of beam and diffuse horizontal radiation is known as the global horizontal radiation. The difference between the global horizontal radiation detected on the surface of the Earth and that at the upper limit of the atmosphere represents the reflected or diffused radiation by the atmosphere.

The direct and diffuse radiation carry information in both horizontal plane to the surface and plane oriented perpendicular to the rays. As such, direct radiation can be measured by a horizontal plane on the surface (Direct Horizontal Irradiance - DHI) or perpendicular to the rays (Direct Normal Irradiance - DNI). Similarly, diffuse radiation can be split into diffuse horizontal and diffuse normal irradiances depending on the orientation of the collecting plane. However, reflected radiation off the Earth's surface can be detected only by a plane oriented perpendicularly to the rays. Table 2.1 summarizes the main classifications of radiation components incident on a surface.

For the case of radiation received by tilted surfaces, the total solar irradiance is an important component. The total solar irradiance accounts for the temporal variability in the instantaneous energy generated by the Sun [13]. Continuous monitoring of the total solar irradiance has revealed its highly variable nature on the temporal scale which is attributed to the 27-day rotation cycle and the 11-year solar cycle [14].

The equations relating the global horizontal radiation (GHI), direct beam radiation at normal incidence (DNI) and the diffuse radiation on a horizontal surface (DFHI) are shown below:

$$GHI = DNI\cos(\phi) + DFHI \tag{2.1}$$

$$GHI = DNI\sin(h) + DFHI \tag{2.2}$$

where ϕ is the solar zenith angle (°) and h is the elevation (°).

Also, the equation relating the Global Normal Irradiance (GNI), Direct Normal Irradiance (DNI), Diffuse Normal Irradiance (DFNI) and Reflected Normal Irradiance (RNI) is stated as follows:

$$GNI = DNI + DFNI + RNI \tag{2.3}$$

The equation relating the GHI, direct horizontal irradiance (DHI) and DFHI is:

$$GHI = DHI + DFHI \tag{2.4}$$

Global solar radiation is uncontestably the most vital and complete solar radiation component since it provides the total solar availability in both spatial and temporal scales [15]. For this reason, emphasis is laid on the global horizontal irradiation parameter for the assessments conducted, as presented in the upcoming chapters.

Table 2.1 Classification of radiation components after interaction with the atmosphere

Radiation component	Definition	Abbreviation
Direct Horizontal Irradiance	Component of radiation received on a flat horizontal plane that originates from the Sun's disc	DHI
Direct Normal Irradiance	Component of radiation detected on a surface held normal to the rays coming from the Sun's disc	DNI
Diffuse Horizontal Irradiance	Component of radiation received on a flat horizontal plane and which originates from scattered light by atmospheric molecules and particles	DFHI
Diffuse Normal Irradiance	Component of radiation that has undergone scattering by atmospheric molecules and particles and received on the plane oriented perpendicular to the rays coming directly from the Sun's disc	DFNI
Reflected Normal Irradiance	Component of radiation that has undergone reflection from the Earth's surface and detected on the plane oriented perpendicular to the rays coming directly from the Sun's disc	RNI

2.5 Photovoltaic Effect

Components of solar radiation reaching a photovoltaic plane has the potential of being converted into electricity. The process of generating electric current from solar radiation in a photovoltaic cell is known as photovoltaic effect.

A solar cell is made up of two different types of semiconductors, one being the p-type and the other one being the n-type. When these two semiconductors are joined together, a p-n junction is formed, as illustrated in Fig. 2.4. As electrons transit to the p-side (positively charged) and holes transit to the n-side (negatively charged), an electric field is set up in the region of the p-n junction. When photons of light of suitable wavelengths are incident on the solar cell, energy is transmitted to electrons in the p-n junction, resulting in electron transition to a higher energy state (conduction band) and giving them freedom of movement [16]. This process leaves a hole in the valence band in the initial position of the electron. Due to the electric field generated

Fig. 2.4 The working principle of a solar cell (Reused with permission from [17] Open Access under a CC BY 4.0 license, https://creativecommons.org/licenses/by/4.0/)

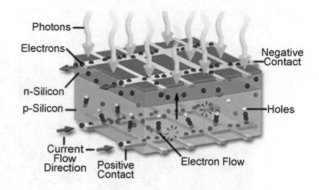

by the p-n junction, the freed electrons are forced to move to the n-side whilst holes are attracted to the p-side. Electron movement in one direction and 'hole movement' in the opposite direction generates a current in the cell.

2.6 Photovoltaic Solar Energy

The photovoltaic effect described in Sect. 2.5 gives rise to the generation of electricity in a photovoltaic cell. The quantity of derived electricity from solar photovoltaic panels is influenced by the exposure and intensity of incident radiation. The global solar radiation is a parameter that is very often used to theoretically estimate the electricity generated from photovoltaic panels. It is given as follows [18]:

$$E = G_s.K.\eta_{pv} \tag{2.5}$$

where E (kWh/m^2 month) represents the monthly mean value of daily electric current generated per unit area; G_s (kWh/m^2 month) denotes the monthly mean daily irradiation; K represents the performance ratio accounting for losses in the system; and η_{pv} represents the conversion efficiency of wafer-based silicon modules. For a more detailed account of the latter two terms, the reader is invited to consult classical textbooks such as Twidell and Weir (2015) [2].

2.7 Measurement of Solar Parameters

Two of the most important parameters for solar energy potential analysis include solar radiation and sunshine duration. On-site measurements of those parameters are usually performed to estimate the the technical potential of a site for solar energy yield. Several instruments are generally used to measure these two parameters in solar radiation and sunshine duration.

Fig. 2.5 Instruments for measuring solar parameters: **a** Kipp and Zonen pyranometer (Reused with permission from [19] Open Access under a CC BY 4.0 license, https://creativecommons.org/licenses/by/4.0/); **b** Pyrheliometer (Reused with permission from [20] Open Access under a CC BY 4.0 license, https://creativecommons.org/licenses/by/4.0/); **c** Campbell-Stokes sunshine duration recorder (Reused with permission from [21] Open Access under a CC BY-SA 3.0 license, https://creativecommons.org/licenses/by-sa/3.0/); **d** Climate monitoring satellite (Reused with permission from [22] Open Access under a public domain license, https://creativecommons.org/publicdomain/zero/1.0/)

The most popular instruments for the measurement of solar radiation include pyranometers and pyrheliometers. The Kipp and Zonen pyranometer (Fig. 2.5a) consists of a thermopile concealed in a glass hemisphere which uses the thermal property of solar radiation to generally measure the global horizontal irradiance. A pyrheliometer (Fig. 2.5b), on the other hand, consists of a collimating tube that is able to measure diffused or direct radiation separately through an in-built mechanism. The sunshine duration parameter is generally measured using a Campbell-Stokes recorder (Fig. 2.5c) which operates under the principle that light rays incident on a glass sphere would focus the radiation into a spot to burn a calibrated card.

Due to long-term maintenance issues and costs incurred with these on-site measuring instruments, geospatial solar energy assessments favour satellite measurements. Climate monitoring satellite technology (Fig. 2.5d) has matured significantly during the past decades and is capable of measuring high resolution solar radiation and sunshine duration parameters through mounted and calibrated sensors. The sensors installed on board of the satellite circumvent the geospatial limitation of point measurements taken by on-site instruments through the measurement of a wider geospatial range. This attribute of satellite measurements makes it particularly useful for geospatial solar energy assessments.

2.8 Spatio-Temporal Variations in Insolation

The spatial and temporal variations in solar irradiance are principally caused by cloud cover processes and seasonal and latitudinal variations. We provide below descriptive statements of how the solar radiation climate is influenced by these factors.

2.8.1 Seasonal, Latitudinal, and Daily Variations

Latitudinal variations in solar radiation occur due to the fact that incoming rays hit the surface nearly perpendicular at the equator and tangential near the poles. Less energy is received at the poles since the same quantity of solar radiation is spread out over much larger distances away from the equator [23]. Moreover, at higher latitudes, incoming solar radiation has to travel through larger atmospheric thickness, experiencing more absorption and reflection than near the equator where the rays penetrate perpendicularly through the atmospheric band [24].

The duration of sunshine also influences latitudinal variations in irradiance. Owing to the tilt of 23.5° between the ecliptic plane and the polar axis of the Earth, regions found at different latitudes of the globe experience a progression of seasons characterised by different extents throughout the year [24]. This mechanism gives rise to different daily irradiance clear sky trends for the same time instant at different latitudes. The resulting uneven variations in net solar insolation over the globe drive the circulation of the atmosphere and ocean, resulting in winds and ocean currents. The rotation of the Earth about its axis gives rise to diurnal variations in insolation which peaks at solar noon and falls to zero after sunset and before sunrise.

2.8.2 Cloud Cover and Topographic Influence

The morphology of clouds varies significantly on both horizontal and vertical scales. The variations are attributed partly to the circulation pattern of the atmosphere and partly to the dispersion of oceans and continents and their multitude and fluctuating sources of condensation nuclei [25]. Clouds regulate the energy equilibrium of the Earth by constantly monitoring the incoming shortwave and outgoing longwave solar radiation. Low level clouds have a cooling effect on the planet since they are thicker and reflect an appreciable amount of extraterrestrial solar radiation, while high thin cirrus clouds have a warming effect due to partially filtering incoming radiation and trapping some of the longwave radiation reflected by Earth's surface [26]. Marine stratocumulus clouds situated on the low levels of the atmosphere highly influence the radiation balance due to their enormous spatial extent, temporal perseverance and substantial reflectance of insolation [25].

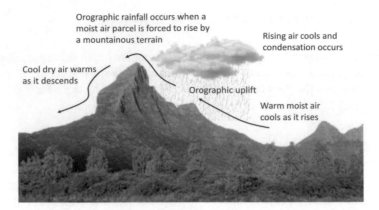

Fig. 2.6 Orographic cloud formation process

The mechanism of solar radiation absorption by clouds is dictated by the optical depth, albedo and phase function of the cloud, in addition to the reflectivity of the surface beneath and the distribution of water vapour in the cloud environment [25]. Water clouds have been found to absorb up to 15% to 20% of solar radiation, with the most absorption occurring from thick clouds possessing massive cloud droplets, an elevated Sun and little water vapour above it [27]. Clouds are found to significantly influence the solar radiation climate on both spatial and temporal scales and the relationship between topography and cloud cover dictates the spatial distribution of insolation. Clouds are also formed as warm moist air is forced to rise due to a mountainous terrain, resulting in a reduction of incoming solar radiation on the windward side of the mountain where rainfall clouds are formed, as illustrated in Fig. 2.6. Rainfall is highest for the windward slopes for large mountain ranges, whilst for smaller hills more precipitation occurs near the crest [28]. The leeward side has relatively higher insolation than the windward side as most precipitation occurs on the windward flank of the mountain, leaving journeying clouds on the leeward side stripped of its moisture content and exposed to the more penetrating to incoming solar radiation.

References

1. Wiki (2012) Diagram of the sun. https://commons.wikimedia.org/wiki/File:Sun_poster.svg. Cited 8 August 2021
2. Twidell J, Weir T (2015) Renewable energy resources. Routledge
3. Inglis M (2015) The sun, our nearest star. in astrophysics is easy! Springer, pp 139–146
4. Khullar D (2014) India: a comprehensive geography. Kalyani Publishers
5. Rottman GJ, Woods T, George V (2007) The solar radiation and climate experiment (SORCE): mission description and early results. Springer Science and Business Media
6. Trenberth KE, Fasullo JT, Kiehl J (2009) Earth's global energy budget. Bull Am Meteorol Soc 90(3):311–324

7. Goswami DY, Kreith F (2007) Handbook of energy efficiency and renewable energy. CRC Press
8. Tchouankap JD, Nguimdo LA (2019) Parameterized transmittance model for atmospheric and surface solar radiations. Atmos Clim Sci 10(1):81–99
9. Barrett EC (2013) Introduction to environmental remote sensing. Routledge
10. Solomon S, Portmann RW, Sanders RW, Daniel JS (1998) Absorption of solar radiation by water vapor, oxygen, and related collision pairs in the Earth's atmosphere. J Geophys Res Atmos 103(D4):3847–3858
11. Wiki (2013) Solar spectrum. https://commons.wikimedia.org/wiki/File:Solar_spectrum_en.svg. Cited 16 August 2021
12. Mohan M (2002) Current developments in atomic, molecular, and chemical physics with applications. Springer Science and Business Media
13. Gueymard CA (2018) A reevaluation of the solar constant based on a 42-year total solar irradiance time series and a reconciliation of spaceborne observations. Solar Energy 168:2–9
14. Frohlich C, Lean J (2004) Solar radiative output and its variability: evidence and mechanisms. Astron Astrophys Rev 12(4):273–320
15. Jain LC, Behera HS, Mandal JK, Mohapatra DP (2014) Computational intelligence in data mining-volume, vol 31. Springer, pp 11–21
16. Gray JL (2003) The physics of the solar cell. Handb Photovolt Sci Eng 2:82–128
17. Tabassum M, Aziz Jahan MA, Rahman MM, Sadik MN (2017) Design and development of maximum power point tracker for solar module using microcontroller. Open Access Libr J 4(05):1
18. JICA (2009) Guideline for application of photovoltaic power generation system. In: Department of energy - renewable energy management bureau
19. Wiki (2012) Typical pyranometer, for measurment of global solar radiation. https://commons.wikimedia.org/wiki/File:SR20_pyranometer_1.jpg. Cited 21 November 2021
20. Wiki (2012) Typical pyrheliometer, for measurement of direct solar radiation. https://commons.wikimedia.org/wiki/File:DR01_pyrheliometer_1.jpg. Cited 21 November 2021
21. Wiki (2002) Heliograph. https://commons.wikimedia.org/wiki/File:Heliograph_wendelstein_2002_00a.jpg. Cited 21 November 2021
22. Wiki (2006) EUMETSAT Meteosat model. https://commons.wikimedia.org/wiki/File:EUMETSAT_Meteosat_model.jpg. Cited 21 November 2021
23. Berner EK, Berner RA (2012) Global environment: water, air, and geochemical cycles. Princeton University Press
24. Stevens AP (2010) Introduction to the basic drivers of climate. Nature education knowledge
25. Hobbs PV (1993) Aerosol-cloud-climate interactions. Academic Press
26. Hilgenkamp K (2005) Environmental health: ecological perspectives. Jones and bartlett learning
27. Twomey S (1976) Computations of the absorption of solar radiation by clouds. J Atmos Sci 33(6):1087–1091
28. Roe GH (2005) Orographic precipitation. Annu Rev Earth Planet Sci 33:645–671

Chapter 3
Geospatial Modelling of Solar Radiation Climate

Abstract Proper land planning and management in terms of agro-suitability of crops and adequate placement of solar photovoltaic and thermal technologies require a knowledge of how the solar insolation parameter varies on spatial and temporal scales. In this chapter, we explore methodologies including satellite remote sensing, Numerical Weather Prediction (NWP) model, and regression analysis that could enable countries to map the spatio-temporal variations in solar radiation. These techniques would offer researchers a route to the proper mapping of the solar resource potential for effective policy decision making.

3.1 Solar Resource Assessment and Optimization

Solar resource assessment involves the characterization of the insolation parameter within a specified domain over a generally long period of time. Due to the inter-disciplinary importance of solar radiation measurements in fields spanning from architecture, agriculture, forestry, ecology, hydrology, meteorology, and energy, proper assessments of the solar resource potential of countries can have wide-ranging benefits and applications [1]. Optimizing the solar resources based on accurate solar radiation assessments may help maximize yield per unit area which may entail significant economic benefits. With ambitious targets set by policy makers to boost renewable energy production, improve agricultural yield, and conserve ecologically-sensitive areas, performing adequate country-wide solar resource assessment studies is a key endeavour. Insolation maps derived from solar energy assessments can also be handy to renewable energy investors, tourists, climatologists and engineers, amongst others.

Solar resource assessments are preferably conducted using a network of pyranometers that is properly distributed at ground stations. The recordings are subsequently processed and interpolated to delineate the spatial variations of the solar resource potential of the region. However, as stated by Nonnenmacher et al. (2014) [2], pyranometers are costly to maintain and measurements are confined to relatively small spatial coverage. To circumvent this problem, researchers worldwide have employed satellite-based measurements to assess the resource potentials of various microclimates [3]. Statistical empirical regression models have also been developed to simulate the insolation curves of distinct regions based on available climate measurements, thereby enabling solar resource assessments to be performed [4]. The emergence of Numerical Weather Prediction (NWP) models has also provided a route towards the characterization of the solar resource potentials of various jurisdictions, and also enabled predictions of the spatio-temporal variations in insolation to be performed [5].

This chapter aims to provide the knowledge on how to conduct a proper solar resource assessment for a given study area. The application of satellite-based, NWP model, and statistical regression techniques are explored to offer researchers a grounding in geospatial data analysis and guide them towards resource optimization.

3.2 Satellite-Based Resource Assessment

Satellite data are usually sought for solar resource assessments owing to their wide spatial coverage. This attribute of satellite images makes them superior to ground-based measurements which record values within a perimeter of the sensor's placement [6]. Additionally, captured images are archived, thereby offering a way to analyze the time evolution of the spatial variations of the climatic parameter. The main limitation of satellite images, though, is the image resolution for climatic parameters which is relatively coarse in general (maximum resolution of 3 km) to conduct a refined analysis for small island states. Figure 3.1 delineates the spatial variations in insolation in India for July 2020 using NASA Earth Observation (NEO) satellite.

Satellite-derived insolation dataset generally employs an algorithm to determine the total amount of radiation reaching Earth under clear-sky conditions. This estimation requires a knowledge on physics modelling that employs the solar radiation values outside of the atmosphere alongside other atmospheric parameters including aerosols optical depth and precipitable water vapor. The algorithm then proceeds by including the effects of additional parameters such as cloud conditions, as characterized by cloud properties derived from satellite image pixels. The cloud cover attenuation factor together with other attenuation parameters dictate the amount of insolation reaching the Earth's surface.

Useful solar resource databases to conduct resource potential assessments include SARAH-E and CLARA-A2, both administered by EUMETSAT's CM SAF [7]. Another solar radiation database includes NASA's Earth Observation (NEO) satellite images which are of sufficiently high resolution on spatial scales and contain relatively long historical data to conduct refined analyses on spatio-temporal scales. SolarGIS provides a high resolution monthly mean annual insolation map of regions of interest, with the limitation of not having archived datasets. Some PV design software such as PVsyst and HOMER also incorporate solar resource databases which could be used to perform assessment studies.

Fig. 3.1 Solar insolation map of India for July 2020 sourced from NASA Earth Observation (NEO) satellite

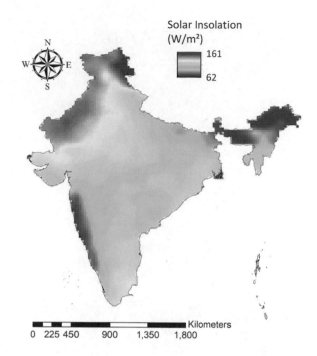

3.3 Weather Research Forecasting (WRF)

WRF is a Numerical Weather Prediction (NWP) system that is tailored to perform climate assessment and forecasting. High spatial and temporal resolution solar resource assessments may be performed using WRF-solar which is designed to provide a representation of the aerosol-cloud-radiation system. A dynamical solver is used to solve the dynamic and thermodynamic equations governing atmospheric circulations. WRF-solar incorporates an aerosol optical property parameter, enabling the representation of the evolving aerosol optical property [9]. An improvement of the aerosol interactions with cloud processes is enabled through the Thompson aerosol-aware microphysics scheme [8]. This scheme enables the differentiation between hygroscopic and non-hygroscopic aerosols. The combination of the hygroscopic and non-hygroscopic aerosol number concentrations is used to determine the aerosol optical depth, which represents the attenuation of the incoming direct beam radiation by aerosols at any wavelength. Characteristics of the incoming solar radiation are thereafter derived.

Numerous solar resource potential studies worldwide have been conducted using WRF. For instance, Ruiz-Arias (2015) [1] mapped the spatial variations in mean global horizontal irradiation for the years 2003–2012 in continental Spain and the Balearic Islands using the WRF-NWP model, as shown in Fig. 3.2. WRF offers the advantage of dynamical downscaling to simulate the physical processes at meso and micro scales, thus enabling high resolution assessments relevant to island scales.

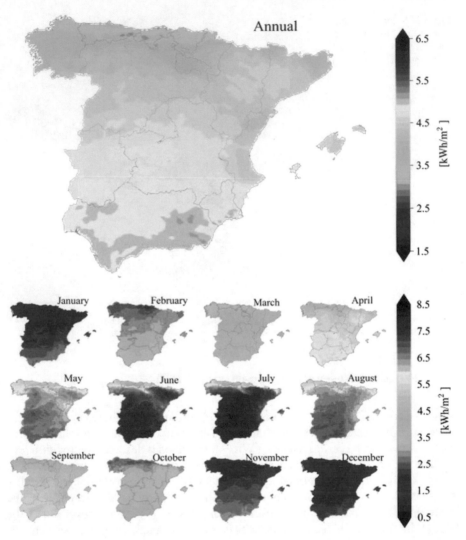

Fig. 3.2 Annual and monthly mean global horizontal irradiation values for the period 2003 to 2012 in continental Spain and the Balearic Islands using WRF simulations (Reused with permission from [1] Copyright (2015) (Elsevier))

3.4 Regression Analysis

Regression analysis is used to analyse the connection between one dependent variable and one or more multiple independent variables [10]. It attempts to study the functional relationship between the variables and thereby provide a technique for prediction or forecasting. The popular regression techniques are: simple linear regression, multiple linear regression, conventional non-linear regression, k-nearest neighbour nonparametric model and logistic regression model [11]. However, since most solar radiation models developed in literature are theoretically based on linear regression, the following lines are devoted to the mathematical foundation underlying this technique.

We consider the simple regression function where y is the outcome, x some explanatory variable, α_0 and α_1 are model parameters:

$$y = \alpha_0 + \alpha_1 x \tag{3.1}$$

Linear regression aims to fit a line through a set of input and output data such as to minimize the summation of absolute errors of fitting for n observations.

$$S^2 = \sum_{i=1}^{n} \varepsilon_i^2 = \sum_{i=1}^{n} (y_i - \alpha_0 - \alpha_1 x_i)^2 \tag{3.2}$$

Through the minimization of S^2, the equations giving the optimum model parameters are given below [12]:

$$\frac{\partial S^2}{\partial \alpha_0} = \frac{\partial S^2}{\partial \alpha_1} = 0 \tag{3.3}$$

The equations lead to the following form [12]:

$$-2 \sum_{i=1}^{n} (y_i - \hat{\alpha}_0 - \hat{\alpha}_1 x_i) = -2 \sum_{i=1}^{n} (y_i - \hat{\alpha}_0 - \hat{\alpha}_1 x_i) x_i = 0 \tag{3.4}$$

This results in:

$$\alpha_1 = \frac{n \sum xy - \sum x \sum y}{n \sum x^2 - (\sum x)^2} \tag{3.5}$$

where $\bar{y}$ and $\bar{x}$ represent the average over n observations of y and x variables.

Besides simple linear regression, other regression models employed to improve accuracy are quadratic, cubic and exponential variants among others.

Multiple linear regression is just an extension of linear regression in the form of multivariate method. For the case of q independent variables [12]:

$$y = \alpha_0 + \alpha_1 x_1 + \alpha_2 x_2 + \ldots \alpha_q x_q \tag{3.6}$$

3.5   Extraterrestrial Solar Radiation

Regression models investigated in this chapter make use of the extraterrestrial irradiation values, H_0, to derive the global horizontal irradiation [13]. The extraterrestrial solar irradiance is derived as shown below according to Besharat et al. [14]. The monthly values of extraterrestrial solar irradiation is then obtained by summing the daily H_0 values derived from Eq. 3.7 over monthly timespan. These are thereafter used in the solar radiation models (described in the following subsection) to estimate the global solar irradiation values at distinct locations.

$$H_0 = \frac{24 \times 3600 G_{sc}}{\pi} \times \left(1 + 0.033 \cos \frac{360n}{365}\right) \left(\cos \phi \cos \delta \sin \omega_s + \frac{\pi \omega_s}{180} \sin \phi \sin \delta\right)$$

(3.7)

where G_{sc} represents the solar constant (1370 W/m^2), n denotes the day of the year, ϕ is the latitude of the site considered (°), ω_s represents the hour angle of the Sun at sunset (°) whilst δ represents the declination (°).

The declination, δ, is derived from [14]:

$$\delta = 23.45 \sin \left[\frac{360}{365} (284 + n)\right]$$

(3.8)

The sunset hour angle, ω_s, is given by [14]:

$$\omega_s = \cos^{-1} \left[-\tan(\delta) \tan(\phi)\right]$$

(3.9)

The day length, S_0, is obtained from [14]:

$$S_0 = \frac{2}{15} \omega_s$$

(3.10)

The insolation detected on the surface of the Earth depends on various climatological variables. The insolation parameter may be related to other climatological variables through regression equations. The following subsection elaborates on model estimations of the insolation parameter using regression models based on climatological parameters.

3.6   Solar Radiation Models

This subsection elaborates on model estimations of the insolation parameter. Many solar radiation models have been developed based on meteorological parameters which have been empirically tested in relation to the variations in solar radiation. Empirical models may be grouped into four main types as described below:

- Sunshine-based models
- Temperature-based models
- Cloud-based models
- Hybrid-parameter-based models.

Table 3.1 gives a description of each model based on relevant literature studies.

3.6.1 Sunshine-Based Models

Several regression equations have been formulated based on the sunshine duration parameter. Some of them are described below:

(1) Angstrom-Prescott model

Initially proposed by Angstrom (1924), the regression model was later simplified by Prescott (1940) through the inclusion of the extraterrestrial solar radiation, H_0 for the clear sky radiation parameter. The Angstrom-Prescott (1940) model [16] is described below [14]:

$$\frac{H}{H_0} = a + b\left(\frac{S}{S_0}\right) \tag{3.11}$$

where H represents the GHI value to be estimated ($MJ/m^2 day$), S is the recorded sunshine hour data ($hours$), a and b represent regression coefficients.

(2) Almorox and Hontoria model

Almorox and Hontoria (2004) [17] used an exponential relation to relate the clearness index to the sunshine ratio for the Spanish territory. The equation is given as follows [14]:

$$\frac{H}{H_0} = a + b\exp\left(\frac{S}{S_0}\right) \tag{3.12}$$

where the sunshine ratio (S/S_0) is exponentially tested against the clearness index (H/H_0).

3.6.2 Temperature-Based Models

The existing relationship between insolation and temperature is founded on the fact that the radiation absorbed by the atmosphere, causes the latter to heat up. This heating takes place due to convection occurring in the layers of the atmosphere, followed

Table 3.1 Types of global solar radiation models and description

Model group	Description	General form
Sunshine-based	Most of the models employed for determining the monthly mean daily global horizontal irradiation use sunshine ratio $\left(\frac{S}{S_0}\right)$ as a basis [14]. Consequently, the sunshine ratio is a function of the ratio of mean daily global horizontal irradiation to the corresponding value on a completely clear day $\frac{H}{H_0}$. Many researchers around the globe have tried to exploit the relationship between these two parameters through the implementation of simple linear, quadratic, cubic, logarithmic, exponential and multiple linear regression techniques. The optimum relationship between these two parameters to achieve good estimates of global solar radiation are found to vary from country to country	a $\frac{H}{H_0} = f\left(\frac{S}{S_0}\right)$
Temperature-based	Temperature is one of the most affordable meteorological parameter which does not necessitate regular maintenance and calibration, in addition to being easily collected and retrieved. As such, implementation of solar radiation models based on commonly available daily minimum, maximum and average air temperature measurements have been widely exploited in several countries. Temperature-based models assume that the ratio of mean daily global horizontal irradiation to the corresponding value on a completely clear day is directly linked to the difference in maximum and minimum temperatures [14]. Nonetheless, several other factors have been observed to influence this difference, among which lies cloudiness, humidity, latitude, altitude, morphology and distance to water bodies [15]	b $\frac{H}{H_0} = f(T)$
Cloud-based	Clouds and their accompanying climatic patterns are observed to significantly restrict the availability of solar radiation on the surface of the Earth [14]. As mentioned previously, clouds cause scattering, diffusion and reflection of incoming solar radiation and dissipates much of the energy coming from the Sun. A number of models have been developed which relates the ratio of average daily global horizontal irradiation to the corresponding value on a clear sky day with mean total cloud cover measurements and cloud factor. However, one major limitation of cloud-based models is that contrary to sunshine-based and temperature-based models, cloud observations are difficult to acquire and relies on satellite and remote sensing measurements which are costly and not readily affordable	c $\frac{H}{H_0} = f(C, CF)$
Hybrid-parameter-based	The increasing trend nowadays in global solar radiation modelling is the implementation of hybrid-parameter-based models which relies on an amalgamation of several meteorological parameters which are observed to be linked with changes in global solar radiation. Several researchers around the globe have tried to obtain a definite set of regression meteorological parameters that best explain the variations in global solar radiation. However, this set is found to vary from location to location due to the distinct microclimate systems where the models are being implemented. The most popular hybrid-parameter-based models relate the ratio of mean daily global horizontal irradiation to the corresponding value on a completely clear sky day with a combination of sunshine ratio, maximum, minimum and average temperatures, cloud cover, cloud factor, relative humidity or precipitation among others	d $\frac{H}{H_0} = f\left(\frac{S}{S_0}, T, \ldots\right)$

[a] Function comprising of sunshine duration S and day length S_0
[b] Function comprising of temperature (includes maximum, minimum and average temperatures)
[c] Function comprising of cloud cover C and cloud factor CF
[d] Function dependent on a mixture of several meteorological variables

by conduction from the upper molecules of the atmosphere, downwards to those near the Earth's surface. Examples of temperature-based models include:

(1) Hargreaves model

A fairly simple relationship was proposed by Hargreaves and Samani [6] linking solar insolation and temperature difference based on a power regression model. It is given as follows [14]:

$$\frac{H}{H_0} = a\,(T_{max} - T_{min})^{0.5} \tag{3.13}$$

where T_{max} is the maximum value of temperature for the day (°C) and T_{min} is the minimum temperature value for the same day (°C).

(2) Pandey and Katiyar model

Pandey and Katiyar (2010) [18] proposed a temperature regression model for India using a third degree temperature-based correlation with solar insolation. It is represented as shown below [14]:

$$\frac{H}{H_0} = a + b\left(\frac{T_{max}}{T_{min}}\right) + c\left(\frac{T_{max}}{T_{min}}\right)^2 + d\left(\frac{T_{max}}{T_{min}}\right)^3 \tag{3.14}$$

3.6.3 Cloud-Based Models

The cloud factor is regarded as one of the most determining parameter that dictates the insolation reaching the Earth's surface. Cloudiness level influences the distribution and dissipation of incoming insolation that penetrates the atmosphere on its trajectory downwards to the surface of the Earth. Founded on the relationship that exists between the ratio of insolation received on the Earth's surface to the extraterrestrial solar radiation, and cloud interception of solar radiation, several cloud-based models have been formulated.

(1) Black model

Black [19] formulated a quadratic equation linking the global horizontal irradiation to the cloud factor, as follows [14]:

$$\frac{H}{H_0} = 0.803 - 0.340C - 0.458C^2 \tag{3.15}$$

where C denotes the total cloud cover (octa) during the day.

(2) Badescu model

Badescu [20] developed linear, quadratic and cubic regression models relating the insolation on the horizontal surface to the cloud cover, as shown below [14]:

$$\frac{H}{H_0} = a + bC \tag{3.16}$$

$$\frac{H}{H_0} = a + bC + cC^2 \tag{3.17}$$

$$\frac{H}{H_0} = a + bC + cC^2 + dC^3 \tag{3.18}$$

3.6.4 Hybrid-Parameter-Based Models

Some regression models have been developed based on an amalgamation of various climatological parameters and are termed hybrid-parameter-based models. They include, amongst others:

(1) Sayigh Universal Formula

The Sayigh Universal Formula [21] is based on a hybrid-parameter model comprising of sunshine hour, relative humidity (R), and maximum temperature values as follows [4]:

$$H = FNK \times \exp\left[\phi_r\left(\left(\frac{S}{S_0}\right) - \left(\frac{R}{15}\right) - \frac{1}{T_{max}}\right)\right] \tag{3.19}$$

where F represents the transformation coefficient (0.01163) to convert from cal/cm^2 day to kWh/m^2 day. ϕ_r denotes the latitude (rad).

Model parameters N and K are determined using [4]:

$$N = 1.7 - 0.458\phi_r \tag{3.20}$$

$$K = 100\left(\frac{0.2S_0}{(1 + 0.1\phi)} + \psi_{i,j}\cos\phi\right) \tag{3.21}$$

where $\psi_{i,j}$ represents a factor acquired from relative humidity in which $i = 1, 2$ or 3 is based on the site relative humidity value ($i = 1$ for $R < 65\%$, $i = 2$ for $R > 70\%$ and $i = 3$ for $65\% \leq R \leq 70\%$), whilst the month is denoted by $j = 1, 2, 3, \ldots, 12$.

(2) Swartman and Ogunlade model

Swartman and Ogunlade [22] formulated a simple linear regression model linking the insolation value on a horizontal surface to the average relative humidity of the site and ratio of sunshine duration and day length, as portrayed in the equation below [14]:

$$\frac{H}{H_0} = a + b\left(\frac{S}{S_0}\right) + cR \tag{3.22}$$

3.7 Mapping Spatially Interpolated Insolation Measurements

The regression models described in this chapter may be applied to climatological parameters recorded by weather stations and interpolated to estimate the spatial variations of the insolation parameter. Hence solar resource potential analysis may be conducted using regression equations based on climatologically-available parameters, thereby offering a cost-effective route to solar resource mapping as compared to using an expensive network of pyranometers.

The material below provides a case study where the Sayigh Universal model, as described in Sect. 3.6.4, has been employed to determine the solar resource potential of Mauritius Island.

Yearly mean daily global solar radiation records, as derived using the Sayigh Universal Formula, and interpolated are presented in Fig. 3.3 for the island of Mauritius. Clear spatial distributions of insolation are observed which range from 17.8 MJ/m² day to 14.2 MJ/m² day. The maximum values tend to occur in the north and western regions while minimum values are observed in the southern part of the central plateau and is principally due to cloud concentrations. The geospatial variations in insolation is attributed to the distinct microclimate regimes influenced by the complex topography of the island. The maritime tropical air mass brought by the persistent South-East Trade Wind throughout the year is the main influence inhibiting cloud formation and consequently rainfall on windward upper plains of the central plateau, as described in Sect. 2.8.2.

Monthly average daily insolation maps of Mauritius for the months of January to December derived using the same Sayigh Universal Formula is shown in Figs. 3.4 and 3.5. The month of November records on average the highest insolation, peaking at 20.15 MJ/m² day while the lowest value on average was detected for the month

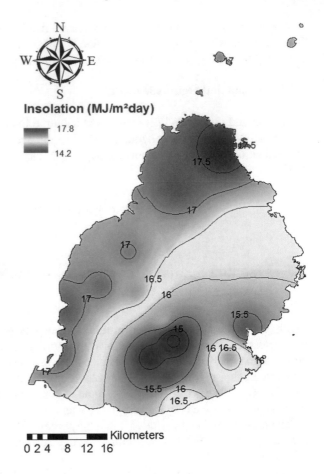

Fig. 3.3 Yearly average insolation map for Mauritius derived using the Sayigh Universal Formula

of June corresponding to 11.70 MJ/m^2 day. This is attributed to high sun angles in the summer month of November, whilst being spread over a much larger area due to low sun angles for the winter months of June and July.

3.8 Relevance of Insolation Maps for Land Management

The assessment and mapping of solar resource potential should be seen as a classic public good, where investment in resource potential analysis can yield long-term and significant financial, social and environmental benefits [23]. Countries investing in solar resource assessments can reap benefits across numerous sectors, including the energy, agriculture and climate sectors, amongst others. Knowledge of the solar resource potential on spatial and temporal scales is of utmost importance in the effective land resource management. Countries faced with limited land constraints

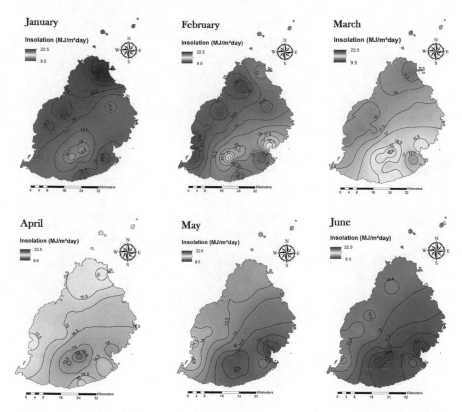

Fig. 3.4   Insolation maps of Mauritius for the months of January to June as derived using the Sayigh Universal Formula

due to climate change and the increasing pace of urbanization find the necessity to optimize on land resources to meet the growing needs of their population.

Solar resource planning may be useful in the agricultural sector to determine the agro-suitability of crops to maximize yield; energy sector to optimize on solar energy capacity; architectural sector to improve on thermal comfort. Consequently, the techniques presented in this chapter, which includes: (1) Satellite-based resource assessment; (2) Weather Research Forecasting; (3) Regression solar models, may support policy makers in making informed decisions pertaining to effective land management. In addition to being useful in the strategies of countries to optimize on solar energy resources, insolation maps will be handy to renewable energy investors, tourists, climatologists and engineers, amongst others.

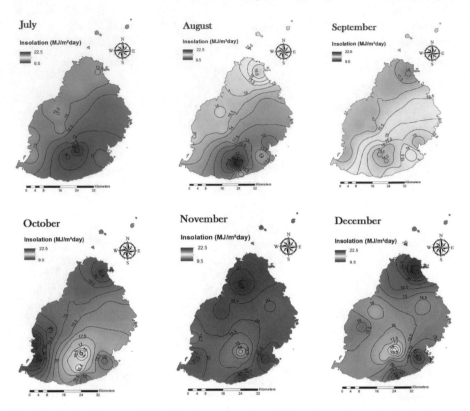

Fig. 3.5 Insolation maps of Mauritius for the months of July to December as derived using the Sayigh Universal Formula

References

1. Ruiz-Arias JA, Quesada-Ruiz S, Fernandez EF, Grueymard CA (2015) Optimal combination of gridded and ground-observed solar radiation data for regional solar resource assessment. Solar Energy 112:411–424
2. Nonnenmacher L, Kaur A, Coimbra CF (2014) Verification of the SUNY direct normal irradiance model with ground measurements. Solar Energy 99:246–258
3. Cano D, Monget JM, Albuisson M, Guillard H, Regas N, Wald L (1986) A method for the determination of the global solar radiation from meteorological satellite data. Solar Energy 37(1):31–39
4. Doorga JR, Rughooputh SD, Boojhawon R (2019) Modelling the global solar radiation climate of Mauritius using regression techniques. Renew Energy 131:861–878
5. Linares-RodrÃnguez A, Ruiz-Arias JA, Pozo-Vazquez D, Tovar-Pescador J (2011) Generation of synthetic daily global solar radiation data based on ERA-Interim reanalysis and artificial neural networks. Energy 36(8):5356–5365
6. Harsarapama AP, Aryani DR, Rachmansyah D (2020) Open-source satellite-derived solar resource databases comparison and validation for Indonesia. J Renew Energy
7. Huld T, Muller R, Gracia-Amillo A, Pfeifroth U, Trentmann J (2017) Surface solar radiation data set Heliosat. Meteosat-East (SARAH-E) 1

8. Thompson G, Eidhammer T (2014) A study of aerosol impacts on clouds and precipitation development in a large winter cyclone. J Atmos Sci 71(10):3636–3658
9. Gamarro H, Gonzalez JE, Ortiz LE (2019) On the assessment of a numerical weather prediction model for solar photovoltaic power forecasts in cities. J Energy Resour Tech 141(6)
10. Bungert M (2012) Termination of price wars: a signaling approach. Springer Science and Business Media
11. Araghinejad S (2013) Data-driven modeling: using MATLAB in water resources and environmental engineering. Springer Science and Business Media, p 67
12. Xuanxuan Z (2018) Multivariate linear regression analysis on online image study for IoT. Cogn Syst Res 52:312–316
13. Duffie JA, Beckman WA (2013) Solar engineering of thermal processes. Wiley
14. Besharat F, Dehghan AA, Faghih AR (2013) Empirical models for estimating global solar radiation: a review and case study. Renew Sustain Energy Rev 21:798–821
15. Allen RG (1997) Self-calibrating method for estimating solar radiation from air temperature. J Hydrol Eng 2(2):56–67
16. Prescott JA (1940) Evaporation from a water surface in relation to solar radiation. Trans Roy Soc S, pp 114–118
17. Almorox JY, Hontoria CJEC (2004) Global solar radiation estimation using sunshine duration in Spain. Energy Convers Manag 45(9–10):1529–1535
18. Pandey CK, Katiyar AK (2010) Temperature base correlation for the estimation of global solar radiation on horizontal surface. Int J Energy Environ 1(4):737–744
19. Black JN (1956) The distribution of solar radiation over the earth's surface. Archiv fur Meteorologie, Geophysik und Bioklimatologie 7(2):165–189
20. Badescu V (1999) Correlations to estimate monthly mean daily solar global irradiation: application to Romania. Energy 24(10):883–893
21. Sayigh AAM (1977) Estimation of total radiation intensity-universal formula. Trans Am Geophys Union 58(8):817–817
22. Swartman RK, Ogunlade O (1967) Solar radiation estimates from common parameters. Solar Energy 11(3–4):170–172
23. World Bank (2016) Assessing and mapping renewable energy resources. https://documents1.worldbank.org/curated/en/317661469501375609/pdf/107219-ESM-P131926-PUBLIC.pdf. Cited 8 August 2021

Chapter 4
Geolocating Optimum Sites for Solar Farms

Abstract Identification of optimum locations for the placements of solar photovoltaic power plants necessitates the consideration of multiple factors, ranging from climatic suitability, technical appropriateness of the land and the legal conforming use of the chosen site. In this chapter, we propose a framework that would enable countries to efficiently harness solar energy through the optimal location of solar farms whilst accounting for social, technical, legal, environmental and climatological factors.

4.1 Utility-Scale Solar Photovoltaic Power Plants

A sharp reduction of about 85% in the cost of photovoltaic modules over the past decade, attributed to the attainment of higher module efficiencies, has transformed the global energy landscape. Utility-scale solar PV has benefited from the favourable economies of scale brought by this cost advantage. The IRENA predicts that global solar PV capacity would increase six fold by 2030 and eighteen fold by 2050 to reach 2,840 GW and 8,519 GW, respectively, from the 480 GW level reported in 2018 [1]. A larger share (~60%) of that exponential growth in total solar PV capacity by 2050 would come from utility-scale PV, whilst the value of installed utility-scale PV plants in 2019 approximated 62% of the cumulative installed photovoltaic capacity [2]. This expansion in utility-scale solar PV was accompanied by a gradual reduction in the levelized cost of unsubsidized utility-scale solar PV which is presently lower than coal power plants and even lower than combined gas cycle plants once subsidies are factored [3]. The cost-competitiveness of utility-scale solar PV as compared to conventional fossil fuel sources encourages its progressive penetration in the global power sector, thereby offering a low-carbon route to development.

Due to significant land requirements (Fig. 4.1), careful planning of solar farms is primordial to optimize solar energy and ensure long-term economic viability. Optimum sites for solar farm placements are dictated by physical factors such as adequate land slopes, technical requirement associated with the proximity to distribution power lines, climatological condition ensuring the abundance of solar energy resource, whilst complying to policy regulations and other spatial constraints [4].

Fig. 4.1 Aerial photographs of the **a** 20 MW Selmer solar farm, Tennessee (Reused with permission from [5] Open Access under a CC BY 4.0 license, https://creativecommons.org/licenses/by/4.0/), and the **b** 16.5 MW Oxford solar farm, Massachusetts (Reused with permission from [6] Open Access under a public domain license, https://creativecommons.org/publicdomain/zero/1.0/)

The advancement of geospatial modelling through GIS has made it possible to identify ideal sites for solar farm constructions through a fairly straightforward route, saving both time and effort as compared to performing in-situ measurements over large spatial extents. GIS facilitates the automated elimination of many non-viable locations, revealing sites that are propitious for exploitation.

This chapter reveals the approach that would enable researchers and energy analysts worldwide to geolocate optimum sites for constructing solar power plants.

4.2   Constraint Regions for Solar Farm Constructions

A fundamental step during the planning phase to determine ideal places for the construction of solar farms is the identification of spatial constraints. Restrictions can either be in the form of legal limitations prohibiting construction within a predefined boundary or socio-technical restraints due to the nature and use of the terrain. A sample set of the main spatial constraints used in solar farm planning (utility-scale) in Mauritius is depicted in Fig. 4.2.

Some areas are protected by strict legal frameworks and are therefore unfeasible for exploitation. Countries promulgate strict laws and legislations to avoid the loss of habitat and prevent the depletion of ecological values [7]. Nature reserves are often identified and excluded from the analysis to determine optimum solar farm sites. World Heritage sites, used to typify the cultural inheritance of a country, are also protected by strict legal frameworks and are removed [8]. Another category which restricts the construction of solar farms are socio-technical factors associated with unexploitable land slopes and the presence of settlement areas. Land slopes of greater than 10% are considered unfeasible for solar farm construction due to the extensive land-levelling works that need to be performed to accommodate the photovoltaic racking system so as to limit damage from soil erosion occurring on steep slopes

Fig. 4.2 Map of Mauritius delineating the location of spatial constraints (legal and socio-technical) in the construction of utility-scale solar PV power plants

■ Settlement area
☐ World Heritage Site
▨ Airport area
▨ Slope > 10%
■ Water body
☐ Nature Reserve

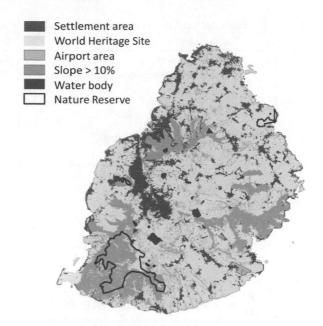

[9]. Also, due to significant land requirements for constructing solar farms, built-up areas are often excluded from the analysis.

Selection of constraints depends on the scale of investigation. At country to continental levels, the constraints shown in Fig. 4.2 would be sufficient. However, for a more refined analysis at the regional level, additional constraints such as parks, monuments, religious sites, wetlands, and helipads that would usually pop up on the surface analysed, need to be identified and excluded.

4.3   Influential Factors Dictating Solar Farm Placements

The choice of location for the construction of photovoltaic power plants is influenced by several factors that complicates the selection problem. Three main categories comprising of climatological, positional and geomorphological factors are extensively used in literature studies to determine optimum sites for solar farm placements. Table 4.1 summarizes the influential factors employed in case studies worldwide to determine ideal sites for solar farms.

Among the set of factors used for solar farm siting, the most popular and influential ones used in literature are: Solar radiation, temperature, slope, aspect and proximity to electric power cables. The identification of a site witnessing high insolation, adequate temperature, appropriate land slope and orientation, whilst being close to high power transmission lines would offer the necessary conditions to optimize on solar energy for electricity generation.

Table 4.1 Criteria and weightage employed to determine optimum solar farm sites

Country	Criteria and weight	Reference
Morocco	· Solar radiation (42%) · Land surface temperature (22%) · Slope (11%) · Land use (5%) · Aspect (7%) · Distance to urban area (7%) · Distance to road (6%)	Tahri et al. (2019) [10]
Brazil	· Solar radiation (23%) · Slope (10%) · Agrological capacity (16%) · Land use (13%) · Proximity to substation (27%) · Proximity to road (7%) · Proximity to urban areas (4%)	Rediske et al. (2020) [11]
England	· Solar radiation (48.9%) · Distance from historical sites (6.5%) · Distance from residential areas (4.9%) · Distance from wildlife designations (6.9%) · Distance from electric network (25.9%) · Distance from transport network (6.9%)	Watson and Hudson (2015) [12]
Iran	· Solar radiation (24.8%) · Slope (4.6%) · Distance from grid lines (15.2%) · Distance from road networks (11.4%) · Distance from settlement areas (8.5%) · Elevation (6.2%) · Relative humidity (1.9%) · Cloudy days (2.5%) · Air temperature (19.9%) · Land use (3.4%) · Dusty days (1.6%)	Mokarram et al. (2020) [13]

Solar radiation is the most important factor in the determination of ideal sites for solar farm constructions since the power generated by photovoltaic panels is proportional to the intensity of incident radiation. In order to generate electricity efficiently, PV panels must be placed in a region having the favourable temperature climate since laboratory experiments have revealed that high temperatures usually result in an associated decrease in energy production [14].

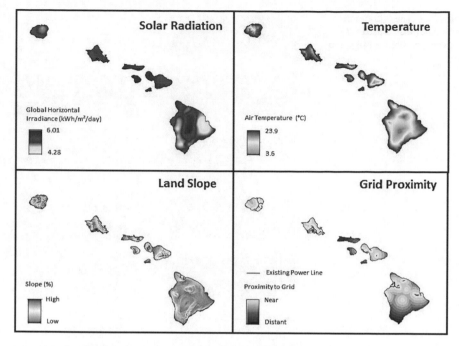

Fig. 4.3 Influential factors that dictate solar farm placements in Hawaii

Besides climatic influences, land specification is another important category in the determination of optimum sites due to the substantial spatial requirements for the construction of solar power plants. Flat lands are usually prioritized in order to reduce the costs in extensive land-levelling work on non-uniform terrains. Also, to optimize on sunshine exposure, having the right PV panel orientation with respect to sun rays would contribute in maximizing electricity generation.

An equally important factor in solar power plant planning is the nearness of the site to power transmission cables. The use of the current infrastructure instead of building new ones cut down on project capital costs and reduces transmission losses due to electricity being transmitted over long distances to finally being injected into the existing grid transmission system [15] (Fig. 4.3).

4.4   GIS-Based MCDA Technique

GIS-based Multi-Criteria Decision Analysis (MCDA) is a general framework for assisting complex decision-making scenarios using multiple geospatial criteria based on value judgements. The main procedures and tools involved in conducting a GIS-based MCDA model are illustrated in Fig. 4.4.

Fig. 4.4 The GIS-based
MCDA process used for
solar farm site identification

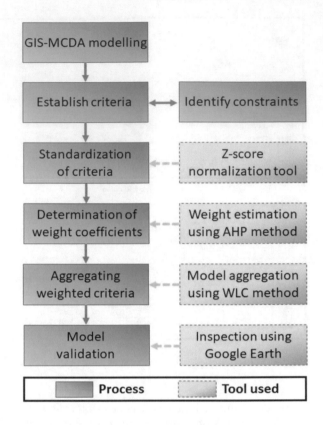

The process begins with the identification of spatial constraints and factors which dictate the placement of solar power plants. The criteria are then standardized to adjust the factors on a common scale. The Z-score normalization technique is used as it helps retain the data structure with low distortion from the mean [16]. Criteria weights are thereafter determined using the Analytical Hierarchy Process (AHP). The standardized and weighted criteria are then combined using the Weighted Linear Combination (WLC) technique to generate a map layer illustrating favourable and unfavourable sites for project implementation. In the final stage, the optimum sites are verified through inspection by analysing satellite images.

The following sub-sections elaborate on the theoretical background and reasoning underlying the main procedures that would help guide the identification of propitious sites for solar farm constructions using a GIS-based MCDA approach.

4.4.1 Z-Score Normalization

Standardization of criteria is an important stage in the GIS-based MCDA model in order to bring all factors influencing solar farm placements on a comparable scale to aid comparison. The Z-score normalization method is used in the current study and gives an indication of the distance of a point in the data sample from its geometrical mean [17]. It is given by the following equation:

$$Z - score = \frac{x_i - \mu}{\sigma} \tag{4.1}$$

where x_i is the value of the criteria x for the ith area, μ is the data mean and σ is the standard deviation of the distribution.

4.4.2 Analytical Hierarchy Process

AHP is a framework of logical thinking enabled through the decomposition of a decision-making problem and subsequent arrangement of perceptions and judgements into a hierarchy of decision priorities that would influence decision making process [18]. Through this approach, a complex problem is decomposed into simpler and more manageable sub-components based on reasoning sub-goals. This technique has been extensively used in GIS-based MCDA models to identify groundwater recharge sites [19], optimum solar farm sites [20], and landslide prone areas [21] amongst others.

The AHP technique is used in the current study to acquire priority weights for criteria influencing solar farm placements. The procedures involved in AHP are:

Step 1: Definition of the problem.

Step 2: Problem structuring through a proposed framework comprised of related criteria influencing the problem.

Step 3: Construction of a decision matrix based on pairwise comparisons of matrix elements.

Step 4: Determination of priority weights associated to each criterion and evaluation of the maximum eigenvalue, consistency index, and consistency ratio of the decision matrix.

The comparison scale shown below is employed in AHP, wherein a value of one indicates comparable importance between two criteria whilst a value of nine indicates the highest order of comparison between the two.

4.4.3 Weighted Linear Combination

The WLC technique falls under the field of bivariate statistical analysis and is a fairly straightforward approach employed in combinatorial analysis whereby a bivariate discriminant function is employed for quantification [22]. For solar farm site identification, the WLC process is performed on the reclassified (using Z-score normalization) and weighted (using AHP) criteria to combine them through a summative linear function. The outcome of this combinatorial process is a map layer delineating favourable and unfavourable sites for solar farm construction. The WLC method is given as follows:

$$WLC_i = \sum_{j=1}^{n} W_j x_{ij} \qquad (4.2)$$

where WLC_i represents the combined suitability estimate for region i, W_j is the criteria weight j, x_{ij} is the normalized value of region i for factor j and n represents the number of criteria.

4.5 Application of the Model for Sites Identification

The current section elaborates how the GIS-based MCDA technique is applied in order to identify suitable sites for solar farm constructions. We take the case of two island states situated in the Pacific Ocean (Hawaii) and Indian Ocean (Mauritius) which are experiencing land limitations in the face of increasing urbanization and development. The approach can be broadened in scale to encompass larger areas, including continental regions.

4.5.1 Case of Hawaii (Pacific Ocean)

The weights attributed to the influential criteria in the combinatorial process for determining optimum solar farm sites, as discussed in Sect. 4.3, are estimated using AHP technique. The criteria deemed important for solar farm planning are as follows:

- Solar radiation (Z_1)
- Proximity to electric power cables (Z_2)
- Slope (Z_3)
- Aspect (Z_4)
- Temperature (Z_5).

Based on pairwise comparisons between factors, the decision matrix shown below was generated and criteria weights were obtained, as shown in Table 4.3.

Table 4.2 Saaty comparison scale employed in AHP

Intensity of weight	Definition	Implication
1	Equal importance	Two criteria share same contribution to objectives
3	Of moderately high importance	One criteria is marginally preferred over another
5	Of strongly high importance	One criteria is strongly preferred over another
7	Of very strongly high importance	One criteria is highly preferred over another
9	Extremely high importance	One criteria is extremely more important than another
2, 4, 6, 8	Intermediate values	Used to represent compromise between priorities
	Reciprocals of above intensity of weights	If criteria i has a non-zero value allocated in comparison to criteria j, then j has the inverse value

Table 4.3 Decision matrix and weights based on a pairwise combination among 5 factors

	z_1	z_2	z_3	z_4	z_5	Weight (%)
z_1	1	3.00	5.00	6.00	8.00	50.4
z_2	0.33	1	3.00	4.00	6.00	25.2
z_3	0.20	0.33	1	3.00	5.00	13.6
z_4	0.17	0.25	0.33	1	3.00	7.1
z_5	0.12	0.17	0.20	0.33	1	3.7

Ten pairwise comparisons were performed among the factors based on the Saaty scale presented in Table 4.2. The resulting decision matrix derived is based on a principle eigenvector, with eigenvalue equivalent to 5.269. The consistency ratio acquired is 6%. Being lower than the 10% threshold value, the consistency ratio indicates that the pairwise comparisons have been consistently made.

The factors were thereafter standardized and combined by applying the correct weights to influential criteria. The resulting map delineating favourable and unfavourable sites for solar farm placements in Hawaiian Islands is shown in Fig. 4.5. Prohibited areas for exploitation are also illustrated in the map to guide solar power plant planning.

With a population of 1.4 million, Hawaii relies on imported petroleum and coal from mainland states to satisfy its energy requirement. The country relies heavily on oil among the 50 US states, and derives approximately 90% of its primary energy requirement from the latter [23]. The highly volatile prices of imported fossil fuels renders the Hawaiian economy particularly vulnerable. In fact, the average electricity

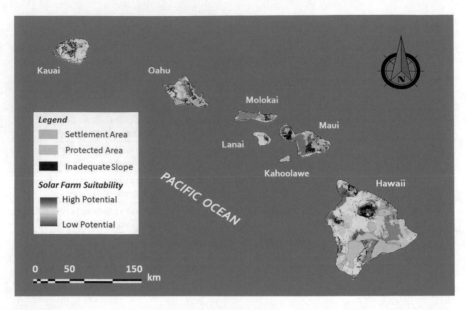

Fig. 4.5 Map of Hawaii's main islands showing favourable and unfavourable sites for solar farm constructions

price in Hawaii has been reported to be twice as high as the average price in the United States [24]. The geographical location of the islands near the equator ensures high solar energy potential and offers the economic prospect of diversifying the energy mix through investments in solar energy facilities. Strategic investments in solar energy could yield economic benefits in the long term via cost savings made through the reduced importation of fossil fuels. However, owing to the ecological richness of this country, development needs to follow a planned route so as to minimize the impacts on the environment and to preserve biodiversity.

Although renewable energy generation in Hawaii is dominated by distributed solar energy system, with an installed capacity of 913 GWh, momentum has been shifting to utility-scale solar energy projects, with the green light being given to solar farm constructions in recent years. This impetus for change has been brought about by significant cost reductions in solar photovoltaic panels as economies of scale tend to favour utility-scale projects over distributed PV installations. The utility company approved the construction of 7 solar power plants with a total capacity of 260 MW in 2018, whilst in 2019, 6 additional projects totalling 247 MW were endorsed [25].

The map shown in Fig. 4.5 can be used to guide planning of solar farms so as to minimize land requirements whilst maximizing on solar energy harness. The relatively higher potential of Maui, the north-western region of Hawaii (Big Island), and the western coasts of Kauai and Oahu in the archipelago can be observed. These regions benefit from high insolation, with global horizontal irradiation attaining $6\,kWh/m^2/day$. Additionally, the closeness of these highly suitable sites to the power

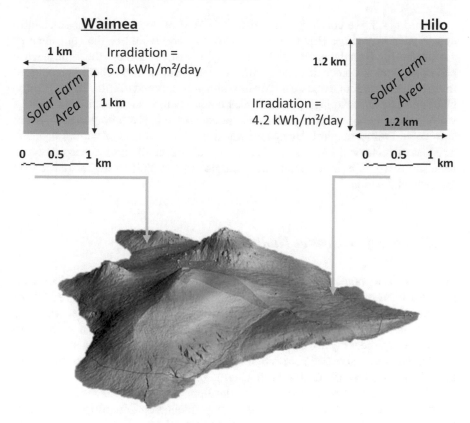

Fig. 4.6 Solar farm area required in Waimea and Hilo to achieve the same generation potential in Hawaii (Big Island)

transmission network, adequate land slopes and favourable temperature regimes enhance the overall appeal of these land parcels for solar power plant constructions.

Solar farm placement in the higher potential area could cut down on land requirements thereby limiting environmental impacts. To illustrate this, we take the contrasting case of Waimea and Hilo, located respectively in the high and low potential regions of Hawaii (Big Island), as shown in Fig. 4.6.

The electric energy derived from solar farm can be obtained using [26]:

$$GP = SR \times CA \times AF \times \eta \tag{4.3}$$

where GP denotes the electric energy potential (kWh/m^2/day), SR is the daily mean global solar irradiation (kWh/m^2/day), CA is the area required for solar farm construction (m^2), AF is a constant which indicates the fraction of land that can be used for the installation of PV panels ($\sim$0.70), and η represents the solar panel efficiency ($\sim$16%).

From Eq. 4.3, we conclude that a parcel of land of about $1\,km^2$ for solar farm construction in Waimea (Irradiation $= 6\,kWh/m^2/day$) would be the equivalent of $1.4\,km^2$ in Hilo (Irradiation $= 4.2\,kWh/m^2/day$) in order to achieve the same electric power generation potential. Land savings of the order of 30% could be achieved through the construction of a solar farm in Waimea as opposed to Hilo. Consequently, building solar farms in high potential sites could result in significant land savings whilst achieving high electric power generation potential. Exploiting high potential sites could limit ecological damage arising from the clearance of natural habitats. The loss of ecological services provided by the clearance of lands to accommodate solar farms would also be significantly mitigated by exploiting higher rather than low potential regions.

4.5.2 Case of Mauritius (Indian Ocean)

Mauritius has a population of 1.2 million and relies heavily on imported petroleum and coal to meet the local energy demand. The country imports about 77.3% of fuel required for electricity production, mostly petroleum products (diesel oil, heavy fuel oil and kerosene) and coal from India and South Africa [27]. In an effort to increase energy security and decrease reliance on imported fossil fuels, several policies targeting the increased penetration of photovoltaic systems have been formulated, including a 3 MW feed-in-tariff scheme (2011), 5 MW net metering scheme (2015), 4 MW PV rebate scheme (2018), development of a nationwide Grid-Code as well as policy incentives encouraging the implementation of rooftop PV and solar farm projects. There are currently 6 operational solar farms on the island and plans have been made to accommodate an additional 4 solar power plants [28]. In the face of this increasing land requirement, careful planning of solar farm placements to optimize on solar energy resources and minimize land space is therefore a necessity in the island's energy strategy.

From 1995 to 2019, agricultural land has decreased from 46.3% to 34%, forest lands and grazing lands have also decreased from 30.6% to 18%, whilst built-up areas have increased from 19.5% to 33% [29]. During that time lapse, natural habitats in Mauritius were mostly destroyed for settlement and agriculture, resulting in fragmented populations of native biota and a marked reduction in biodiversity [30]. In order to conserve the remaining green spaces on the island, land allocated for development must target brown, biodiverse-deficient areas rather than green, ecologically-rich areas. In the island's quest to meet the rising energy need of the population, it is important that efforts are made to conserve the remaining native terrestrial biodiversity.

Using the GIS-based MCDA technique and combining the standardized and weighted factors listed in Table 4.3, resulted in the solar farm suitability map shown in Fig. 4.7. It can be observed through inspection that the higher potential region is found in the northern plain of Mauritius, with high insolation ($\sim$5 $kWh/m^2/day$), proximity to electric grid network, suitable temperature, and adequate land slope

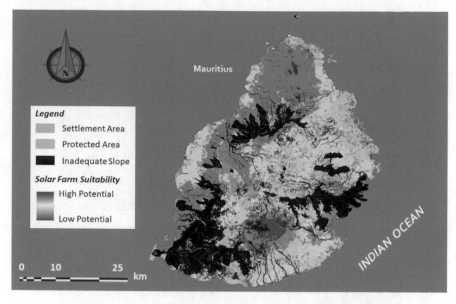

Fig. 4.7 Map of Mauritius showing favourable and unfavourable sites for solar farm constructions

and orientation. For the same electric power generation potential, constructing solar farms in the northern plains could result in significant land savings as opposed to building them on the central uplands where the solar farm potential and insolation are fairly low. In this Anthropocene epoch where the impacts of global warming and climate change are being felt mostly by small island states, it is vital that the management of land resources are made in a thoughtful and strategic manner.

4.6 Concentrated Solar Power (CSP) Systems

CSP systems generate electricity when reflected lights from mirrors are concentrated onto a receiver to convert solar energy into thermal energy that would then operate a heat engine (Fig. 4.8). Global CSP capacity rose by 11% in 2019, reaching 6.2 GW, with Spain leading installation and followed by emerging markets in Israel, Kuwait and France [31]. IRENA has reported reductions in costs of CSP which fell by 46% from 2010 to 2018 and attributed to improved supply chain in China, accrued expertise and competitive auctions [32]. The multi-criteria framework proposed in Fig. 4.4 for utility-scale solar PV farm planning may be extended to encompass CSP farm planning. The list of influential criteria needs to be amended to cater for water availability for cooling CSP plants, as performed by Aly et al. (2017) to geolocate optimum sites for CSP power plants in Tanzania [33]. Nonetheless, the methodology adopted would remain the same as described in the earlier sections of this chapter.

Fig. 4.8 Greenway CSP Mersin Solar Tower (Reused with permission from [34] Open Access under a CC BY 4.0 license, https://creativecommons.org/licenses/by/4.0/)

Fig. 4.9 Illustration of a floating solar system on an irrigation pond in California (Reused with permission from [36] Open Access under a CC BY-SA 3.0 license, https://creativecommons.org/licenses/by-sa/3.0/)

4.7 Floating Photovoltaic Systems

In territorially limited regions, the implementation of floating photovoltaic systems on reservoirs, wetlands, lakes, water treatment plants, and irrigation ponds (Fig. 4.9) may help boost the share of renewable energy in the electricity generation mix. Besides necessitating less terrestrial space, floating photovoltaic systems hold the benefit of decreasing water evaporation from in-land water bodies and increasing electricity yield due to the natural cooling effect of water on solar panels [35]. The placements and spatial extents of inland water bodies also benefit from less structural obstructions from neighbouring buildings and infrastructures that would block sunlight. Floating structures also reduce algae growth and potentially improves water quality [35]. However, countries subjected to stormy climate should limit their deployments to small inland water bodies that would experience wave heights of no greater than 0.5 m. The methodology presented in this chapter could also be used to determine the most appropriate location among all water bodies within a certain territory to accommodate floating solar panels.

4.8 Solar Hybrid Options

Hybrid power systems that combine photovoltaic solar energy with another energy source are referred to as solar hybrid systems. The most popular one being the PV diesel hybrid system which operates on the principle that diesel generators would gradually fill in the gap between the load and the power produced by the photovoltaic system. PV diesel hybrid system is best used in unreliable grid situations with frequent power outages or off-grid cases where electricity generation from photovoltaic systems can be appropriately maximized, resulting in an associated reduction on fossil fuel energy dependence. This would encourage increased PV penetration in countries with low and poor grid capacities. Solar-wind hybrid systems are also increasingly being implemented worldwide.

4.9 Boosting Energy Production and Minimizing Land Space

To improve the climate resilience of countries globally, utility-scale solar photovoltaic power plants are being constructed at a rapid rate, occupying thousands or millions of acres of land. This land use change has serious implications on the wildlife and habitat as power plants are usually fenced, impacting the preying strategy, hiding spots and food availability [37]. The soil is often scraped to bare ground whilst in some cases, the vegetation is frequently mowed in order to prevent vines and grasses intercepting incoming sun rays received by photovoltaic panels. Since there is a direct relationship between the impact on wildlife and biodiversity, the affected wildlife in or around power plants will give rise to a less diverse or practically non-existent ecosystem in case of cleared lands. Moreover, whilst deforestation reduces the land's natural carbon sequestration rate, it is important that land clearance for the construction of solar power plants is kept to a minimum in order to conserve natural carbon stores even at the expense of environmentally sustainable projects.

Countries having land constraints worldwide are faced with the dilemma of having to increase energy supply to meet the rising energy needs of their populations whilst building a climate resilient economy to mitigate the impacts of climate change. Solar energy could provide an economically viable and environmentally robust route for countries, especially those located near tropical and sub-tropical belts, to transit to low carbon economies. As lands are being sought for urban development, agriculture and construction, the sustainable management of land resources is of prime importance, particularly in the global context of climate change where lands are being degraded by the increased intensity and frequency of heavy precipitation and heat stress. It is therefore vital that any form of development involving substantial land usage, such as for the implementation of solar power plants, to be carefully planned in order to optimize on energy harnessed whilst minimizing land requirement. The mitigation of environmental and ecological impacts stemming from the implementation of solar farms would enhance the overall appeal of resorting to utility-scale projects in order to dampen the reliance on fossil fuels.

The GIS-based MCDA approach presented in this chapter offers a route for countries to assess their solar farm suitability potential by taking into consideration legal, technical, climatological and economic factors. We apply the method on the islands of Hawaii and Mauritius which are facing substantial land constraints in the face of increasing urbanization and development. The technique can however be broadened to cover a much larger spatial area depending on the availability of high-resolution data.

4.10 Policy Incentives to Increase Utility-Scale Solar Capacity

Partial or complete tax and import duty exemptions on solar equipment could relieve investors and result in significant cost savings where substantial investments in utility-scale projects are concerned. For utility-scale solar farms, crowd-funding platforms could help meet the necessary financial target for project implementation. Auctions, nonetheless, represent the most popular procurement mechanism for utility-scale solar PV. Financing from multilateral financial institutions and development banks represent a significant share of successfully implemented utility-scale solar projects. However, to allow for the increasing penetration of intermittent solar power in the grid, especially at the utility-scale, investment in Battery Energy Storage Systems (BESS) is of prime importance.

A holistic approach to policy formulation would necessitate that certain standards and norms be respected to avoid proliferation and dumping of low-quality panels. Independent Power Producers (IPPs) should be encouraged to deal with the defective or end-of-life solar equipments through exportation incentives or industrial symbiosis and recycling initiatives.

References

1. IRENA (2019) Future of photovoltaic: deployment, investment, technology, grid integration and socio-economic aspects (A Global Energy Transformation: paper). International Renewable Energy Agency, Abu Dhabi
2. Lindberg O, Birging A, Widen J, Lingfors D (2021) PV park site selection for utility-scale solar guides combining GIS and power flow analysis: a case study on a Swedish municipality. Appl Energy 282:116086
3. Shen W, Chen X, Qiu J, Hayward JA, Sayeef S, Osman P, Meng K, Dong ZY (2020) A comprehensive review of variable renewable energy levelized cost of electricity. Renew Sustain Energy Rev 133:110301
4. Palmer D, Gottschalg R, Betts T (2019) The future scope of large-scale solar in the UK: Site suitability and target analysis. Renew Energy 133:1136–1140
5. Wiki (2015) Selmer solar farm 20MW. Available via Wikimedia Commons. https://commons. wikimedia.org/wiki/File:Selmer_Solar_Farm_20MW.jpg. Cited 25 April 2021
6. Wiki (2017). 16.5 MW Oxford solar farm. U.S. Department of Energy. Available via Wikimedia Commons. https://commons.wikimedia.org/wiki/File:Faria_000108_153922_493466_4578_(35529343313).jpg. Cited 22 April 2021
7. Aydin NY, Kentel E, Duzgun HS (2013) GIS-based site selection methodology for hybrid renewable energy systems: a case study from western Turkey. Energy Convers Manag 70:90–106
8. Leask A, Fyall A (2006) Managing world heritage sites. Elsevier, vol 1, pp 1–20
9. Doorga JR, Rughooputh SD, Boojhawon R (2019) Multi-criteria GIS-based modelling technique for identifying potential solar farm sites: a case study in Mauritius. Renew Energy 133:1201–1219
10. Tahri M, Hakdaoui M, Maanan M (2019) The evaluation of solar farm locations applying geographic information system and multi-criteria decision-making methods: case study in southern morocco. Renew Sustain Energy Rev 51:1354–1362
11. Rediske G, Siluk JCM, Michels L, Rigo PD, Rosa CB, Cugler G (2020) Multi-criteria decision-making model for assessment of large photovoltaic farms in Brazil. Energy 197:117167
12. Watson JJ, Hudson MD (2015) Regional Scale wind farm and solar farm suitability assessment using GIS-assisted multi-criteria evaluation. Landsc Urban Plan 138:20–31
13. Mokarram M, Mokarram MJ, Khosravi MR, Saber A, Rahideh A (2020) Determination of the optimal location for constructing solar photovoltaic farms based on multi-criteria decision system and Dempster-Shafer theory. Sci Rep 10(1):1–17
14. Huld T, Amillo AMG (2015) Estimating PV module performance over large geographical regions: the role of irradiance, air temperature, wind speed and solar spectrum. Energies 8(6):5159–5181
15. Azevedo VWB, Candeias ALB, Tiba C (2017) Location study of solar thermal power plant in the state of Pernambuco using geoprocessing technologies and multiple-criteria analysis. Energies 10(7):1042
16. Talukder B, Hipel WK, vanLoon GW (2017) Developing composite indicators for agricultural sustainability assessment: effect of normalization and aggregation techniques. Resources 6(4):66

17. Misra S, Osogba O, Powers M (2019) Unsupervised outlier detection techniques for well logs and geophysical data. Mach Learn Subsurf Charact 1
18. Saaty TL (2000) Fundamentals of decision making and priority theory with the analytic hierarchy process. RWS publications, p 6
19. Chowdhury A, Jha MK, Chowdary VM (2010) Delineation of groundwater recharge zones and identification of artificial recharge sites in West Medinipur district, West Bengal, using RS, GIS and MCDM techniques. Environ Earth Sci 59(6):1209–1222
20. Colak HE, Memisoglu T, Gercek Y (2020) Optimal site selection for solar photovoltaic (PV) power plants using GIS and AHP: a case study of Malatya Province, Turkey. Renew Energy 149:565–576
21. Jam AS, Mosaffaie J, Sarfaraz F, Shadfar S, Akhtari R (2021) GIS-based landslide susceptibility mapping using hybrid MCDM models. Nat Hazards, pp 1–22
22. Umagandhi R, Kumar AS (2017) Web usage mining techniques and applications across industries. IGI Global, pp 199–222
23. Lee T, Glick MB, Lee JH (2020) Island energy transition: assessing Hawaii's multi-level, policy-driven approach. Renew Sustain Energy Rev 118:109500
24. Arent D, Barnett J, Mosey G, Wise A (2009) The potential of renewable energy to reduce the dependence of the state of hawaii on oil. IEEE, pp 1–11
25. Wikipedia (2021) Solar power in Hawaii. Available via Wikipedia. https://en.wikipedia.org/wiki/Solar_power_in_Hawaii#cite_note-17. Cited 15 April 2021
26. Asakereh A, Soleymani M, Sheikhdavoodi MJ (2017) A GIS-based Fuzzy-AHP method for the evaluation of solar farms locations: case study in Khuzestan province. Iran Solar Energy 155:342–353
27. Surroop D, Raghoo P (2017) Energy landscape in Mauritius. Renew Sustain Energy Rev 73:688–694
28. CR (2019) Mauritius connects 16.3MW solar park to the grid. https://constructionreviewonline.com/news/mauritius-connects-16-3mw-solar-park-to-the-grid/. Cited 5 May 2021
29. Mauree PP (2020) Re-visiting our land use planning strategy in Mauritius. https://www.lemauricien.com/le-mauricien/re-visiting-our-land-use-planning-strategy-in-mauritius/370447/. Cited 5 May 2021
30. Florens FV (2013) Conservation in Mauritius and Rodrigues: challenges and achievements from two ecologically devastated oceanic islands. In: Conservation biology: voices from the tropics, pp 40–50
31. REN21 (2020) Renewables 2020: global status report. https://www.ren21.net/wp-content/uploads/2019/05/gsr_2020_full_report_en.pdf. Cited 3 September 2021
32. IRENA (2019) Renewable power generation costs in 2018. https://www.irena.org/-/media/Files/IRENA/Agency/Publication/2019/May/IRENA_Renewable-Power-Generations-Costs-in-2018.pdf?la=en&hash=99683CDDBC40A729A5F51C20DA7B6C297F794C5D. Cited 3 September 2021
33. Aly A, Jensen SS, Pedersen AB (2017) Solar power potential of Tanzania: identifying CSP and PV hot spots through a GIS multicriteria decision making analysis. Renew Energy 113:159–175
34. Wiki (2015) Mersin CSP field. Available via Wikimedia Commons. https://commons.wikimedia.org/wiki/File:Mersin_CSP_field.jpg. Cited 3 September 2021
35. Sahu A, Yadav N, Sudhakar K (2016) Floating photovoltaic power plant: a review. Renew Sustain Energy Rev 66:815–824
36. Wiki (2018) Floating PV system Far Niente Winery California. Available via Wikimedia Commons. https://commons.wikimedia.org/wiki/File:Floating_PV_system_Far_Niente_Winery_California_2018.jpg. Cited 26 September 2021
37. Turney D, Fthenakis V (2011) Environmental impacts from the installation and operation of large-scale solar power plants. Renew Sustain Energy Rev 15(6):3261–3270

Chapter 5
Optimizing on Rooftop Solar Technologies–Photovoltaic and Solar Water Heating

Abstract Rooftop photovoltaic and solar water heating technologies have enormous potential to decrease reliance on fossil fuel sources for domestic energy needs. Decarbonizing the domestic sector through solar energy technologies holds the promise of a more resilient, reliable, and affordable clean energy future. In this chapter, we explore both solar energy technologies and provide a methodology to estimate the rooftop solar potential at a country scale. We then suggest policy incentives to address potential roadblocks in the development and expansion of domestic solar energy systems.

5.1 Rooftop Solar Energy Potential

Increasing solar penetration at residential and commercial levels are expected at global scales in the upcoming decades. The rooftop PV installation market is predicted to expand at a compound annual growth rate of 11.2% from 2017 to 2023 [1]. Government policies which include net metering and feed-in-tariff schemes have helped to boost the global rooftop PV market. This rise in global rooftop PV installations is further encouraged by global cost reductions in the installation and supply of photovoltaic panels.

Besides the sustainable generation of electricity from rooftop PV systems, the conversion of solar to thermal energies by solar water heaters for domestic and commercial purposes, has helped in the overarching goal for the decarbonization of commercial and residential buildings. The global demand for solar water heaters is predicted to increase at a compound annual growth rate of 6.7% between 2020 to 2026 [2]. The high demand for hot water for residential and commercial buildings (e.g. hotels, hospitals), supported by Government policies, are the key market drivers that contribute in the expansion of the global solar water heater market.

Cities have a fundamental role to play in the integration of distributed renewable energy generation, and examples of progressive cities which are actively involved in the deployment of solar energy technologies, exist worldwide. The city of Barcelona has been engaged in the formulation of energy-related policies, including the adoption of a thermal solar ordinance which demanded buildings to use at least 60% of solar energy [3]. The Polish city of Niepolomice invested 17.3 million euros over the years 2012–2015, with the objective to equip 3,841 households and 32 public buildings with solar panels, solar water heaters and heat pumps, demonstrating that viable and economically attractive alternatives to coal exist [4]. The city of San Francisco was the first major US city to require 15–30% rooftop solar PV or green roofs on all new building rooftops [5]. Through policies enabling the progressive integration of distributed solar energy technologies, Cities hold the power to dampen fossil fuel dependence for energy needs.

Numerous tools are available which allow the determination of the radiation performance on building rooftops. On the micro-level (architectural, urban-scale), the most popular technique is Computer Aided-Design (CAD) which consist of simulating the solar radiation performance based on the Sun's declination and structural obstructions [6]. Examples of micro-scale models include RADIANCE lighting simulation model [7]; TOWNSCOPE II [8]; SOLENE [9]. However, these tools necessitate huge computational resources which limit their application to some extent. For macro-level analysis, GIS-based methods enable the determination of the rooftop solar energy potential at regional and country-scales through the parameterization of the rooftop space by population density. This technique has the benefit of being fairly low in computational intensiveness and is relatively easy to use for technical rooftop solar potential estimations at large scales. The methodology underlying the GIS-based method will be further elaborated in the upcoming sections of this chapter.

5.2   Rooftop Photovoltaic Potential

Countries worldwide have to assess the technical potentials of their energy resources to meet energy and climate targets. The majority of countries lying in tropical and subtropical regions have emphasized on the importance of rooftop PV potential to meet increasing global energy demand. Consequently, the quantification of the rooftop PV potential is a global necessity for the setting up of realistic targets regarding the deployment of rooftop photovoltaic panels (Fig. 5.1). A geospatial analysis for the European Union estimated that about 25% of the electricity consumption of the Member States could be met by rooftop photovoltaic deployment, representing an annual economic potential of 467 TWh [10]. Another study using LiDAR (Light Detection And Ranging) technology estimated that 57% of the 116.9 million residential units in the United States are adequate for rooftop PV technology with the capacity to generate an annual energy potential of 1,000 TWh, representing 75% of residential consumption [11].

Geospatial assessments of rooftop photovoltaic potential at national level is important in order to determine which zones or regions need to be prioritized for the promulgation of policies to promote rooftop PV systems. Priority should be given to regions witnessing high solar radiation and having favourable financial and policy incentives which would reduce the cost of solar electricity. The cost-competitiveness of solar energy with regards to fossil fuel sources would play a fundamental role in the global

Fig. 5.1 Rooftop photovoltaic system (Reused with permission from [12] Open Access under a CC BY-SA 3.0 license, https://creativecommons.org/licenses/by-sa/3.0/)

power sector decarbonisation issue. Besides the environmental benefit of promoting rooftop PV in favourable areas, market trends and dynamics could further drive the global cost of solar energy downwards, which could make it economically advantageous to promote rooftop PV systems in the intermediate and lower solar potential areas. However, for this to occur, the market signal needs to respond positively to technical, policy and economic factors. An alignment of all these contributing factors could help increase the deployment rate of rooftop PV systems worldwide to meet global energy needs.

5.3 Solar Water Heating (SWH) Potential

The continued reliance on coal, oil and natural gas for heating and cooking, has contributed in the increase of carbon dioxide emissions from the building sector which attained around 10 GtCO$_2$ in 2020, representing a share of 28% of the global energy-related carbon dioxide emissions [13]. It is estimated that residential and commercial buildings contribute to around 54.7% and 45.3% of the overall energy consumption in the building sector, respectively, with the highest consumption attributed to space and water heating [14]. Consequently, solar water heaters represent a feasible alternative to decrease the emissions of carbon dioxide from the building sector for water heating. This is principally due to the relatively lower maintenance requirement and ease of operation associated with the installation of solar water heaters. However, the cost-effectiveness of this technology is dependent on the policy incentives provided by governments and the cost of fossil fuels for water heating [15] (Fig. 5.2).

Governments worldwide have implemented a set of policy measures and fiscal incentives aimed at increasing the global uptake of solar water heaters, to be used as an alternative to electric water heaters. China has been leading the adoption of solar water heaters globally, followed by Turkey, India, Brazil, and Germany [17].

Fig. 5.2 Solar water heater
(Reused with permission
from [16] Open Access
under a CC BY 4.0 license,
https://creativecommons.org/
licenses/by/4.0/)

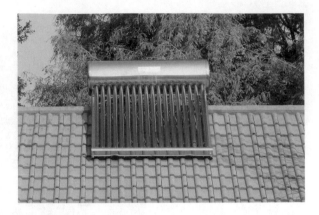

Financial and fiscal incentives have been granted in European countries in the form of subsidies, soft loans and tax credits to encourage the installation of solar water heaters in residential settings [18]. For the small island developing state of Mauritius, a solar water heater scheme involving a subsidy was granted to provide assistance to families, resulting in a reduction in gas consumption for residential water heating of about 786 tons per year whilst savings in electricity was estimated around 4.4 GW annually [19]. Due to the existing potential to further exploit the solar water heating industry worldwide, we investigate in this chapter a methodology to assess its potential, from a geospatial perspective.

5.4   A Geostatistical Approach for Rooftop Potential Assessments

An assessment of the rooftop potential for PV or solar water heater deployments would necessitate that the available space be determined. Determination of the available space for exploitation requires the determination of reduction coefficients, including: (1) Obstructions (C_o) such as water storage tanks or TV antenna; (2) Shadows cast (C_s) by structures or trees; (3) Orientation (C_a) of the roof with regards to the Sun; (4) Racking space (C_r) for rows of PV panels to optimize on incoming insolation. However, owing to the fact that these parameters vary from rooftop to rooftop, especially for different building types (residential, commercial, industrial, educational, and hospitals), a sampling needs to be performed for each building category.

The total rooftop area of a country may be estimated using the linear relationship that exists with population size (see Sect. 5.4.3 below).

Fig. 5.3 Sample rooftop
design of a residential unit

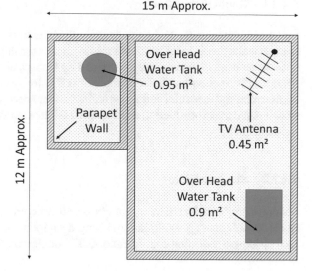

5.4.1 Sampling of Building Rooftops

As an illustration, a sample rooftop design of a residential unit is shown in Fig. 5.3. Obstructions to the deployment of PV panels or solar water heaters are the overhead water tanks, parapet wall and TV antenna with a combined obstruction area of 37.3 m², as estimated through surface area computations. The total rooftop area is estimated at 162 m². The obstruction coefficient is therefore calculated as follows:

$$C_o = 1 - \frac{Obstruction}{Total\ Area} = 1 - \frac{37.3}{162} = 1 - 0.230 = 0.77 \qquad (5.1)$$

Similarly, the shadow coefficient is estimated using the following equation:

$$C_s = 1 - \frac{Shadow\ area}{Total\ Area} \qquad (5.2)$$

Sampling of several rooftop designs for different localities and different building categories and averaging the results would give representative values for the reduction coefficients.

5.4.2 Acquisition of Rooftop Coefficients

Rooftop building coefficients to evaluate the available space for rooftop exploitation are derived using the following methodology.

5.4.2.1 Roof Obstructions

Overhead water tanks, staircase rooms, air extractors, roof signs, chimneys, and parapet walls decrease the available rooftop space for the placements of PV panels or solar water heaters. A sampling performed by Doorga et al. (2021) [20] for Mauritius revealed obstruction coefficient values of 0.895 for residential buildings; 0.936 for commercial and industrial buildings; and 0.954 for specialty buildings. Consequently, an estimated mean obstruction coefficient of 0.928 was acquired for building rooftops in Mauritius.

5.4.2.2 Shading Factor

Shadows projected by tall trees, buildings and structures reduce the available rooftop space for solar energy exploitation. Using a sampling strategy, Doorga et al. (2021) [20] estimated a shading coefficient of 0.959 on the rooftops of Mauritian buildings.

5.4.2.3 Orientation

To determine the aspect coefficient, C_a, Wiginton et al. (2010) [21] proposed the following equation:

$$C_a = r_{flat} \cdot f_{flat} + r_{pitch} \cdot f_{pitch} \qquad (5.3)$$

where r_{flat} denotes the ratio of flat rooftops, r_{pitch} represents the ratio of pitched rooftops, f_{flat} represents the reduction undergone by flat roofs, and f_{pitch} denotes the reduction undergone by pitched roofs.

For the case of Mauritius, Doorga et al. (2021) [20] estimated that flat roofs represent 95% of the building rooftops and therefore undergo no reduction ($f_{flat} = 1$). Half of the pitched roofs, on the other hand, are north-facing and exposed to high insolation throughout the year. The aspect coefficient is therefore estimated as follows:

$$C_a = r_{flat} \cdot f_{flat} + r_{pitch} \cdot f_{pitch} = (0.95 \times 1 + 0.05 \times 0.5) = 0.975 \qquad (5.4)$$

5.4.3 Estimation of Total Rooftop Area

The total rooftop area of a country may be estimated using the linear relationship that exists with population size. A simple linear regression involving total rooftop area and population size may be modelled using available information on building

Fig. 5.4 Building rooftop
extraction at the local
neighborhood scale (Reused
with permission from [21]
Copyright (2010) (Elsevier))

footprint and population at the neighborhood level (Fig. 5.4) and then extrapolated
to municipal, and district-levels. As an illustration, Fig. 5.5 depicts the geospatial
distribution of building footprints across the various districts of Mauritius which
may be related to the population statistics in those districts in an attempt to model
the linear regression. This linear trend may then be extrapolated to estimate the
total rooftop area at the country-scale based on the known population parameter.
An overall rooftop area per capita, $A_{RA/capita}$, of 94.96 m^2/capita was estimated for
Mauritius [20].

5.4.4 Available Rooftop Space for PV and SWH Exploitations

The available rooftop space for solar energy exploitation is related to the available
area per capita, $A_{Solar/capita}$, and the population size. The available area per capita
is derived from the overall rooftop footprint per capita, $A_{RA/capita}$, roof obstructions
(C_o), shading (C_s), and aspect (C_a) as follows:

$$A_{Solar/capita} = A_{RA/capita} \times C_o \times C_s \times C_a \times C_r \qquad (5.5)$$

For the case of Mauritius, the available area per capita is given by:

$$A_{Solar/capita} = 94.96 \times 0.928 \times 0.959 \times 0.975 = 82.4 \text{ m}^2/\text{capita} \qquad (5.6)$$

The total available area for solar energy exploitation (PV or SWH) is then determined as follows:

$$A_{Solar} = A_{Solar/capita} \times Population \qquad (5.7)$$

Fig. 5.5 Geospatial
distribution of building
footprints at the
country-level for Mauritius

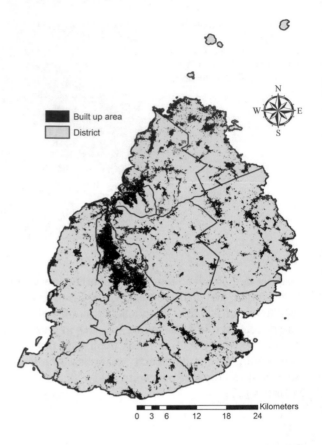

For population size of 1,274,336 (Mauritius), the total available area for solar energy exploitation is given by:

$$A_{Solar} = 82.4 \times 1,274,336 = 105,001,586\,\text{m}^2 \qquad (5.8)$$

Consequently, the total available area for PV and SWH deployments is estimated at 105,001,586 m^2.

A solar farm of 15 MW capacity at La Ferme on the island, constructed on about 340,000 m^2 of land, generated roughly 0.73% of the electricity requirements of the country [22]. With an available rooftop space of about 300 times the size of the solar farm, rooftop energy holds much promise for Mauritius.

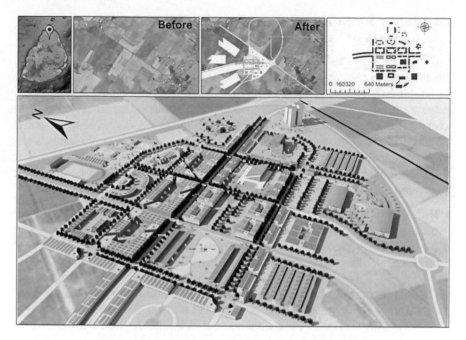

Fig. 5.6 Solar City design to encourage the exploitation of rooftop solar potential (Reused with permission from [20] Copyright (2021) (Springer Nature))

5.5 Solar Urban Planning

Regarded as the powerhouses of economic growth, cities which accommodate more than half of the global population are responsible for the major share of greenhouse gas emissions. Cities worldwide have followed a rather unsustainable development route throughout the last decades with the overwhelming dominance of grey infrastructures and unsustainable building designs. The spatial constraints posed by cities for utility-scale renewable energy integration have inspired a redefinition of urban planning. In this context, solar urban planning is a popular theme that is gaining traction and which seeks to exploit the solar energy potential at the urban scale to attain smart energy cities composed of zero energy buildings [23]. Urban architectures that exploit the available rooftop space for solar technologies integrations need to be encouraged. Doorga et al.(2021) [20] proposed a Solar City design (Fig. 5.6) for the tropical island climate region of Mauritius whereby component buildings fulfil the key requirements of optimizing on solar energy potential through the integration of rooftop solar PV systems (inclined at optimum angle) and solar water heaters. Solar urban planning has an important role to play in achieving SDG 11 to make cities inclusive, safe, resilient and sustainable.

5.6   Policy Incentives to Drive Rooftop PV and SWH Adoption

Countries worldwide, especially small island developing states, need to transit to low-carbon, climate resilient economies in order dampen reliance on the highly volatile and expensive fossil fuels. Countries blessed with abundant solar resources have the potential to make significant savings on energy imports in the domestic sector, which relies extensively on energy-intensive activities encompassing air conditioning, water heating, refrigeration, and transportation. However, the route to low-carbon transition in the domestic sector is plagued with barriers hampering the uptake of rooftop photovoltaic and solar water heaters. These barriers include:

- **Difficulty in accessing start-up capital**: In developing and underdeveloped countries, capital markets lack maturity and banks do not support investments in low-carbon technologies due to offtake risks involved. Limited access to finance is a key barrier in the solar energy industry.

- **Limited institutional capacity**: Installation of rooftop PV or SWH necessitates a conducive environment with an established supply chain. Capacity building to ensure quality service by a trained workforce is limited in some countries due to the cost implications of such an exercise.

- **Sustained policy and regulatory support**: A conducive policy and regulatory framework for upscaling solar technology adoption is of high importance in boosting the solar market. Limited grants, loans, building regulations, energy policy instruments, and tax incentives tend to limit adoption of solar energy technology at the domestic level due to high up-front cost involved.

- **Limited consumer awareness campaigns**: Consumer awareness of solar energy technology purchase options remains relatively low, thereby hindering its adoption and subsequent market expansion. An effective market strategy, built on solar energy technology education and pricing incentives can improve consumer acceptance.

- **Grid readiness**: The ability of the current grid to accept the influx of intermittent solar energy from a network of integrated rooftop PV systems on available roof space poses a technical limitation.

- **Limited private sector involvement**: Besides the strong marketing and communication attributes of the private sector, the latter has the potential to decrease upfront costs of solar energy technologies by assisting the government in its endeavour of boosting renewable energy capacity. Limited public-private partnerships is regarded as an important barrier in the switch to renewable energy technologies.

In order to boost the rooftop photovoltaic and solar water heater markets, aggressive policy actions in the form of tax rebates or other financial support need to be provided. Green loans and long term financing options offered by the private banking sector may help contribute in this key endeavour. Private-public partnership is fundamental in order to spread the upfront costs associated with solar energy technologies, whilst ensuring strong marketing aimed at improving consumer awareness on pricing facilities and benefits of low-carbon energy adoption. It is necessary for governments to invest in capacity building exercises to improve the institutional capacities of countries and provide confidence for investors and consumers.

A conducive policy package comprising of fiscal incentives, regulatory certainty, de-risking strategies, and increased duty on fossil fuels, gas and electric heaters would contribute in increasing solar energy capacity at the domestic level. Building regulations which encourage the adoption of rooftop solar energy technologies need to be reviewed and implemented. Residential-scale rooftop PV systems could be supported through feed-in-tariff and net-metering schemes, provided that they are properly designed and implemented to avoid conflicting policy targets whilst attracting investors without undermining the financial stability of the stakeholders. In some cases, a Renewable Portfolio Standard (RPS) forcing electricity generators, including fossil fuel-based companies, to either invest in solar energy technologies or offset their carbon-intensive process through the acquisition of Solar Renewable Energy Credits (SRECs) from small-scale electricity generators, could help boost the country-wide solar energy market.

References

1. Doshi Y (2017) Rooftop solar photovoltaic (PV) installation market by technology (thin film and crystalline silicon), grid-type (grid connected and off-grid), and end-use (residential and non-residential) - global opportunity analysis and industry forecast, 2017–2023. https://www.alliedmarketresearch.com/rooftop-solar-photovoltaic-PV-installation-market. Cited 5 September 2021
2. ZMR (2020) Solar water heater market: global industry perspective, comprehensive analysis and forecast 2020 - 2026. https://www.researchandmarkets.com/reports/5144134/solar-water-heater-market-global-industry?utm_source=BW&utm_medium=PressRelease&utm_code=xczlh2&utm_campaign=1450818+-+Insights+on+the+Solar+Water+Heater+Global+Market+to+2026+-+Rising+Demand+Solar+Water+Heating+Systems&utm_exec=jamu273prd. Cited 5 September 2021
3. Newman P, Matan A (2013) Green urbanism in Asia: the emerging green tigers. World Scientific
4. CMW (2017) Cities at the forefront of climate action: achieving the Paris climate goals through the effort sharing regulation. https://carbonmarketwatch.org/wp-content/uploads/2017/10/ACHIEVING-THE-PARIS-CLIMATE-GOALS-THROUGH-THE-EFFORT-SHARING-REGULATION.pdf. Cited 6 September 2021
5. Keeley M, Benton Short L (2018) Urban sustainability in the US. Cities Take Action: Springer
6. Desthieux G, Carneiro C, Camponovo R, Ineichen P, Morello E, Boulmier A, Abdennadher N, Dervey S, Ellert C (2018) Solar energy potential assessment on rooftops and facades in large built environments based on lidar data, image processing, and cloud computing. methodological background, application, and validation in geneva (solar cadaster). Front Built Environ 4:14

7. Compagnon R (2004) Solar and daylight availability in the urban fabric. Energy Build 36(4):321–328
8. Teller J, Azar S (2001) Townscope II-A computer system to support solar access decision-making. Solar Energy 70(3):187–200
9. Miguet F, Groleau D (2002) A daylight simulation tool for urban and architectural spaces application to transmitted direct and diffuse light through glazing. Build Environ 37(8–9):833–843
10. Bodis K, Kougias I, Jager-Waldau A, Taylor N, Szabo S (2019) A high-resolution geospatial assessment of the rooftop solar photovoltaic potential in the European Union. Renew Sustain Energy Rev 114:109309
11. Sigrin BO, Mooney ME (2018) Rooftop solar technical potential for low-to-moderate income households in the United States. National Renewable Energy Lab (NREL), Golden, CO (United States)
12. Wiki (2009) Rooftop photovoltaic array. https://commons.wikimedia.org/wiki/File:Rooftop_Photovoltaic_Array.jpg. Cited 12 September 2021
13. UNEP (2020) 2020 global status report for buildings and construction: towards a zero-emission, efficient and resilient buildings and construction sector. https://wedocs.unep.org/bitstream/handle/20.500.11822/34572/GSR_ES.pdf?sequence=3&isAllowed=y. Cited 19 September 2021
14. Jahangiri M, Akinlabi ET, Sichilalu SM (2021) Assessment and modeling of household-scale solar water heater application in zambia: technical, environmental, and energy analysis. Int J Photoenergy
15. Roulleau T, Lloyd CR (2008) International policy issues regarding solar water heating, with a focus on New Zealand. Energy Policy 36(6):1843–1857
16. Wiki (2016) Domestic solar heater in South Africa. https://commons.wikimedia.org/wiki/File:DomesticSolarHeater_SouthAfrica.jpg. Cited 18 September 2021
17. Gautam A, Chamoli S, Kumar A, Singh S (2017) A review on technical improvements, economic feasibility and world scenario of solar water heating system. Renew Sustain Energy Rev 68:541–562
18. Lele SB (2016) Reducing GHG emission by regrouping world population emergence of EUCAn. Water Energy Int 59(8):14–20
19. Businessmega (2013) Solar water heater: 10,000 new beneficiaries in january. https://business.mega.mu/2013/12/14/solar-water-heater-10000-new-beneficiaries-january/. Cited 19 September 2021
20. Doorga JRS, Tannoo R, Rughooputh SD, Boojhawon R (2021) Exploiting the rooftop solar photovoltaic potential of a tropical island state: case of the Mascarene Island of Mauritius. Int J Energy Environ Eng, pp 1–18
21. Wiginton LK, Nguyen HT, Pearce JM (2010) Quantifying rooftop solar photovoltaic potential for regional renewable energy policy. Comput Environ Urban Syst 34(4):345–357
22. CEB (2021) Production facts and figures. https://ceb.mu/our-activities/production-facts-and-figures. Cited 21 September 2021
23. Amado M, Poggi F (2012) Towards solar urban planning: a new step for better energy performance. Energy Procedia 30:1261–1273

Chapter 6
Conclusions

Abstract This chapter provides conclusive remarks on the main theme of this book regarding sustainable land planning to attain energy optimization. The socio-economic benefits associated with increased solar energy penetration are highlighted. We summarize/reflect on the material presented in the earlier chapters and provide recommendations for future work that would enable a more comprehensive view on the topic.

6.1 Land Planning to Approach Energy Optimization

The pace of urbanization and competing land use to satisfy the rising socio-economic needs of the global population are raising concerns on issues of climate change and energy security. The increasing population densities in cities and built infrastructures are making it difficult to redefine the urban landscape, in view of increasing energy efficiency and sustainability. Construction of utility-scale power plants on vast expanse of lands diminishes the available land resources for ecological-environmental spaces and alternative socio-economic uses especially in territorially limited regions. In view of complying with ambitious climate targets established by international environmental treaties whilst minimizing further environmental and ecological damages, a redefinition of sustainable land management in the context of energy security is of absolute necessity.

Planning and design of the built environment should seek to make efficient use of the existing urban footprint. Exploiting rooftop space for integrating solar energy technologies in the urban fabric would help circumvent the geospatial constraint caused by low available land space in territorially limited regions. An assessment of the rooftop potential for solar energy exploitation in Mauritius, as elaborated in Sect. 5.4, revealed a technically feasible rooftop space of 105 km^2 for solar PV and/or solar water heater deployments. The exploitation of this available space would necessitate government intervention though policies and regulations which seek to incentivize the adoption of rooftop solar energy technologies, along with ordinances to allow the progressive penetration of the latter in building designs at the domestic level. To ensure the sustainable land management and in view of minimizing environmental and ecological impacts, priority for solar energy technology deployment should be given to available rooftop spaces.

In the eventuality of the necessary implementation of utility-scale ground-mounted solar energy projects, compact solar farm designs need to be prioritized through their placements in a high solar resource potential areas to ensure maximum yield in minimum space. Mixed-use development which seeks to co-locate solar photovoltaics and agriculture or livestock should be encouraged. The geospatial methodology which identifies spatial constraints and factors for site selection, as described in Chap. 4, needs to be adopted and construction on relatively barren lands with low soil productivity levels should be prioritized. Forested areas, with high vegetation indices need to be avoided due to high value of lands in the global fight against climate change. Furthermore, as most water bodies are state-owned, floating solar farms on dams and irrigation ponds offer promising venues.

6.2   Towards Energy Security and Socio-Economic Welfare

Increasing solar energy penetration has the potential to make a strong contribution towards improving the security of energy supply. The expansion of the solar energy market would also help in reducing negative externalities in the form of environmental and health impacts. Increasing the share of solar energy in the global energy mix would lead to a reduction in fossil fuel energy consumption, thereby contributing to meet the climate goals needed to prevent unprecedented and irreversible climate change effects. Most countries, being signatory to the UN Sustainable Development Goals and the Paris Agreement, are bound to strive towards achieving climate targets. Consequently, efficient solar energy exploitation is pivotal in achieving SDGs 7 (affordable and clean energy), 11 (sustainable cities and communities), 13 (climate action), whilst meeting the goals of the Paris Agreement.

An important socio-economic impact stemming from the expansion of the solar energy market is the creation of employment opportunities which may support small and medium enterprises associated with the manufacture, delivery and installation of solar energy technologies at both local and international levels. Furthermore, reduced dependence on imported fossil fuel resources to meet increasing energy needs could save millions of direct revenues which could be invested in the health, education and employment sectors. In a state of global economic crisis, the right investments in the right direction could spill numerous social benefits to other sectors. Maximizing solar energy penetration would ensure energy security and a pathway to socio-economic welfare.

6.3 Closing Remarks

The general purpose of this book is to provide a guide to help researchers assess the solar resource potential of countries worldwide in view of supporting science-based policy decisions. A step-by-step approach is taken, in which the solar energy processes are first defined and elaborated to equip the reader with an understanding of the solar parameter being explored. We then proceed by elaborating on the methodologies that may be employed to perform solar resource assessments on spatial and temporal scales. Such assessments are pivotal in effective land planning to optimize on agricultural, energy, and thermal comfort needs. In order to guide researchers perform effective solar farm siting analysis, a robustly-tested methodology is proposed which considers climatological, social, technical and economic factors whilst rejecting regions protected by laws or which are technically unfeasible for project implementation. Furthermore, to provide a holistic view on the exploitation of solar energy, a methodology is proposed to assess the technical rooftop potential of buildings for photovoltaic system and solar water heating deployments.

The material presented in this book should equip readers with the following:

- A theoretical foundation governing solar energy processes.
- Knowledge of the tools and techniques used in solar resource assessment.
- Knowledge of the approach used for effective solar farm planning.
- Knowledge of the method used for rooftop solar potential assessment.
- Knowledge of effective energy policies to boost the solar energy market.
- Instil effective terrestrial and inland water bodies planning concepts in order to minimize environmental and ecological impacts.
- Provide innovative solutions to integrate rooftop solar technologies and exploit solar resource potential at the urban scale.

6.4 Recommendations for Future Work

The focus of the materials presented in this book is limited to on-grid solar energy. For a more comprehensive view on the geospatial exploitation of solar energy resources, the scope of the current study should be broadened to encompass off-grid and mini-grid for electrification of regions where access to electricity is lacking. Heat storage for space heating may also need to be reviewed in line with the strategy to decarbonize the global energy mix through solar energy penetration.

Appendix A
Data Sources

The sources of data employed in the current study are varied and diverse while the temporal scales of the measurements differ. Table A.1 below presents a description of the main sources of data. Meteorological data was mainly made available from NASA Earth Observation Satellite. The Revue Agricole et Sucriere de Île Maurice and climatological summaries of the Mauritius Meteorological Services also came handy while modelling the solar resource potential. Important spatial information in Mauritius was handed to us by the Ministry of Housing and Lands (Cartographic Section) and OpenStreetMap. Building footprints acquired from Ministry of Housing and Lands were also useful.

J. Doorga et al., *Geospatial Optimization of Solar Energy*,
SpringerBriefs in Energy, https://doi.org/10.1007/978-3-030-95213-6

Table A.1 Data specifications, as employed in this study

Chapter	Data	Period	Source
Solar radiation modelling (Chap. 3)	Mean monthly values of daily sunshine duration; Global solar radiation; Relative humidity; Temperature (minimum, maximum, mean)	1961–1990; 1994–2003; 2011–2016; 2020	Mauritius Meteorological Services; Revue Agricole et Sucriere de Ile Maurice; NASA Earth Observation satellite
Solar farm site identification (Chap. 4)	Digital elevation model; Slope; Aspect; World heritage sites; Protected Areas; Major settlement areas; High transmission lines; Annual mean daily global solar radiation, temperature	2000–2010; 2018	NASA Earth Observation satellite; Ministry of Housing and Lands; Mauritius Meteorological Services; OpenStreetMap
Rooftop solar technologies (Chap. 5)	Building footprints; Population statistics	2017–2019	Ministry of Housing and Lands; Statistics Mauritius

Appendix B
Analysis Tools and Techniques

A number of different tools and techniques have been employed in this study to successfully assess and model the solar resource potential of countries. An overview of the main tools and techniques used in each chapter is presented in Table B.1. All processing were performed using an Intel(R) Core(TM) i7-8750H operating at 2.20 GHz with a 16 GB central processing unit.

© The Author(s), under exclusive license to Springer Nature Switzerland AG 2022

J. Doorga et al., *Geospatial Optimization of Solar Energy*,
SpringerBriefs in Energy, https://doi.org/10.1007/978-3-030-95213-6

Table B.1 Tools and techniques, as employed in this study

Chapter	Process	Technique	Tool
Solar radiation modelling (Chap. 3)	Regression analysis	Simple linear, quadratic, cubic, logarithmic, exponential, exponent, multiple linear	MATLAB (Version: R2015a)
	Interpolation	Inverse Distance Weighting	ArcMap (Version: 10.3.1)
Photovoltaic resource modelling (Chap. 4)	Trend fitting	Polynomial regression	MATLAB (Version: R2015a)
	Feature digitization	Creation of polygon shapefile	ArcMap (Version: 10.3.1)
	Proximity analysis	Euclidean distance	ArcMap (Version: 10.3.1)
	Interpolation	Inverse Distance Weighting	ArcMap (Version: 10.3.1)
	MCDM	Weighted linear combination	ArcMap (Version: 10.3.1)
	Criteria weightage	AHP	AHP priority calculator
Rooftop photovoltaic assessment (Chap. 5)	Rooftop sampling	Area calculation	ArcMap (Version: 10.3.1)
	Trend fitting	Linear regression	MATLAB (Version: R2015a)
	City and urban planning	Architectural modelling	Google SketchUp (Version: 2008)

Index

Printed in the United States
by Baker & Taylor Publisher Services